FOOD FOR THOUGHT

FOOD FOR THOUGHT

Catholic Insights into the Modified Food Debate

John Perry SJ

NOVALIS

© 2002 Novalis, Saint Paul University, Ottawa, Canada

Cover: Caroline Gagnon
Cover photograph: Eyewire
Layout: Caroline Gagnon

Business Office:
Novalis
49 Front Street East, 2nd Floor
Toronto, Ontario, Canada
M5E 1B3

Phone: 1-800-387-7164 or (416) 363-3303
Fax: 1-800-204-4140 or (416) 363-9409
E-mail: cservice@novalis.ca

National Library of Canada Cataloguing in Publication Data

Perry, John F. (John Frederick), 1946–
    Food for thought : Catholic insights into the modified food debate / John Perry.

Includes bibliographical references.
ISBN 2-89507-241-8

    1. Genetic engineering—Religious aspects–Catholic Church. 2. Genetic
engineering—Moral and ethical aspects. 3. Genetically modified foods–Religious
aspects–Catholic Church. 4. Genetically modified foods–Moral and ethical aspects.
I. Title.

TP248.65.F66P47 2002   261.5'6   C2002-902724-1

Printed in Canada.

We acknowledge the financial support of the Government of Canada through the
Book Publishing Industry Development Program (BPIDP) for our publishing
activities.

10 9 8 7 6 5 4 3 2 1   10 09 08 07 06 05 04 03 02

Contents

This book is dedicated to the late Dr. Baldur Stefansson of the University of Manitoba, who died on January 3, 2002, at the age of 84. A plant breeder who started a new farm industry in Canada and abroad through his development of canola from rapeseed, Dr. Stefannson never personally profited from his discovery. He represents well the "magnificent obsession" of the scientific project of improving the world's food supply, a project shared by those working today with genetically modified organisms.

Acknowledgments

Many people helped the author in the labour of researching and writing this book. He would like to thank: Sheila Watson and her husband, Bill; Michael Stogre, SJ; Rob Allore, SJ; Edward O'Donnell, SJ; Tibor Horvath, SJ; Edward Dunnett; the Robertson Company of Ajax, Ontario; Kay Yee; Steve Perry; Betty Perry; and Kevin Burns.

Foreword

Each day we are confronted with questions about genetically modified foods. We are told by some that these new technologies will usher in a new era of productivity and plenty; others say they will create monumental health and environmental problems, a Pandora's box better left unopened. Whom are we to believe?

Fr. John Perry underlines the influences these technologies can have on the future of the human family. He also clearly outlines who stands to profit from them and who will ultimately pay. This is a clarion call for government, which is responsible for the common good, to play its proper role. The public needs to be informed; this book is a very helpful tool for the lay person to use to learn more.

In this readable book, this contemporary issue, which at first appears to be bewilderingly complex, is clearly explained. Principles of science and morality are laid out in an accessible way, and the subsequent questions become obvious.

Food for Thought is an important book. I hope and pray that John Perry's work will reach a large audience, enabling ordinary men and women to give the issue of genetically modified foods the attention it requires if we are going to build a world of peace and justice.

Most Rev. James Weisgerber
Archbishop of Winnipeg
Chairman, Social Affairs Commission
Canadian Conference of Catholic Bishops

Introduction

"We are what we eat." This piece of folk wisdom speaks to the deep-seated importance of food and what goes into it for the ordinary person. Research on public acceptance of transgenically modified food indicates that consumers are experiencing growing anxiety about its ever-increasing presence in our shopping carts at local food stores.

There are many ways in which this dis-ease, connected to the widespread use of biotechnology intended to improve our food, is expressed. On April 4, 2001, the Chief Provincial Court Judge of Prince Edward Island, Mr. Justice John Douglas, convicted Evan Wade Brown of common assault committed against the Prime Minister of Canada, Jean Chrétien.[1] In August 2000, while the Prime Minister was touring an agricultural exhibition in Prince Edward Island, Mr. Brown, as part of the self-styled PEI Pie Brigade, attacked him with a paper plate filled with whipped cream. The previous year, the RCMP had also placed Mr. Brown under arrest. Mr. Aaron Koleszar, a spokesperson for the group, said that the Pie Brigade had criticized Mr. Chrétien in part because his government "supported and encouraged the biogenetic altering of food."[2]

Many in the biotechnology and scientific community dismissed the antics of members of the Pie Brigade as well as the more violent protests in Seattle in December 1999, in Quebec City in April 2001, and in Genoa a few months later, as behaviour beneath their contempt. However, comparative research conducted by the Canadian opinion polling organization Ipsos Reid, on behalf of the British journal *The Economist,* demonstrates widespread rejection of this technology throughout the world.[3] In the year 2000, two-thirds of Canadians interviewed by Ipsos Reid pollsters said that they would be less likely to buy food that they knew had been genetically manipulated, preferring unmodified food. In Britain, this proportion is even higher: 82 per cent. This is a major reason why the food industry has not supported the labelling of genetically modified (GM) products despite widespread public demand that it do so.

Since 1994, North American farmers have grown more than 3.5 trillion plants that have been subjected to gene splicing.[4] The story of the introduction of this new food technology and the mobilization of popular opposition to it has all the ingredients of a suspense novel that you might purchase while waiting at the checkout counter to pay for your groceries – approximately 60 per cent of which, we are told, have already been genetically altered in one way or another.[5] The plot has heroes, heroines and villains; unexpected developments; high-level political intrigue; organizations bent on world domination; and organizations trying to save the world from destruction.

What is it about this question of food that raises such passion? I am not sure, and this book will provide no real answer. But one idea is simply that food is special. It has both mythic or, if you prefer, theological symbolism, as well as a more prosaic meaning. We care about the story of food biotechnology much more than other stories of commercial practice or malpractice. From the beginning, food has been so fundamentally material, so present at every stage of our cultural history that it is not thought of as symbolic of anything. At the same time, it is fraught with meaning. Roman Catholics have died in battle or have been hanged, drawn and quartered by an executioner

12

for the sake of the truth that the bread and wine of the Eucharistic feast are *really*, and not merely *symbolically*, the Body and Blood of Christ. At the beginning of the rise of modern economic life, food seemed too sacred, too important to us to be handled in a free market. Adam Smith argued against this position in his famous addendum on the grain trade. Today, visit any commodity exchange and you will see that "futures" on soy, canola, wheat and pork prove that Adam Smith was prescient: food has become just one of the facts of nature.

This book represents a modest contribution to the ongoing debate over transgenically modified food. On offer here will simply be some perspectives of the author, a Roman Catholic, as well as some other Christians on this vital question. They share the misgivings expressed in the opinion surveys; therefore the question of why consumers often experience this visceral reaction against gene splicing in their food will be central to the book. Why do people hold this viewpoint? Whence does this reaction come? In general, the book will display two perceptions on the question at once: a deep-seated ambivalence toward the novel food created by biotechnology, along with the hope that the technology itself will help us solve many serious problems in our world. Resources within Catholic tradition and theology will help us resolve our mixed reactions to transgenic modification. At the very least, the reader will have more food for thought after reading this book.

As in many other matters, individual Catholics are free to choose whether or not to develop or use these new food products. They are encouraged to learn as much as they can about the questions and about how theologians and other leaders in the Church of the past have commented on related issues. At the moment, arguments that rely on analogy are the only ones available because there does not yet exist any detailed "official" Catholic teaching on genetically modified organisms (GMOs).

The Teachings of the Church

To date, Pope John Paul II has spoken directly on the issues under discussion just twice. On the more recent occasion, His Holiness reminded his listeners on the Jubilee of Farmers that advances in biotechnology, as they apply to agriculture, cannot be evaluated merely on the basis of "immediate economic interests" but must be examined carefully beforehand through a rigorous scientific and ethical assessment.[6] In both speeches, he has given clear directions for Catholic reflection on the new technology in line with the idea of "solidarity" with the poor and the hungry, and participation of all in its application and use. In October 2000, a collection of essays on the technology was published by the Vatican's Pontifical Academy for Life, which also gave their qualified and careful agreement to biotechnology in general and to genetic modification of food in particular.[7]

Lack of an official position on this question means, among other things, that Catholic theologians can approach the question of whether recent developments in biotechnology are congruent with the best interests of human persons and the flourishing of their communities with intellectual care combined with comparative academic freedom. This is not the case with some other moral issues, such as abortion or non-therapeutic experimentation on early embryos.

The very process of non-therapeutic experimentation on human embryos is judged by the official teaching of the Church to be immoral. By contrast, many Catholics would encourage the efforts of molecular biologists and technicians working in biotechnology. While the research might be commendable, it is the possible outcomes that concern some Catholics. For instance, Pope John Paul II and his Vatican advisers do have misgivings about the issue of justice as it relates to this and similar issues. Given that any technology can be used for good or ill, they ask who will really benefit from the remarkable possibilities that scientific research has opened up to the family of humankind. Specifically, will the poor in Africa, Central and South America, and Asia benefit or not? Will the result merely be that the rich stockholders of the multinational biotechnology companies become richer and the poor to whom they sell their new seeds and other farm products poorer? Because the answer to this question is by no means clear at this point, there is a basic, albeit cautious, openness within some significant sectors of the Catholic community to developments in this field.

And why should Catholic perspectives matter? Indirectly, this question was implied in a recent comment in *Nature Biotechnology*, a journal dedicated to this new technology. Commenting on a press conference held by the Vatican Pontifical Academy for Life (VPAL), the pope's advisers on life-science matters, a writer made the patronizing statement that the VPAL "has decided that it is not against the will of the catholic God to alter the genetic make-up of plants and animals."[8]

The notion of a "catholic God" implies many things at once, all of them mistaken. One would be that the God of Catholics is different from the God of any other believer. This is false. Another implication is that what the papal advisers or, for that matter, the pope himself, think about these issues does not really matter. This also is false. The faith that honours God, whether expressed by a Catholic, Protestant, Muslim or Hindu, will want to know what Catholic perspectives on the GMO question are so that a reasonable and responsible choice can be made in the light of faith.

The theological methodology that will be employed in the GMO question assumes a "faith-informed autonomous ethic" of this type. Since the book is addressed primarily to believers in general, and specifically to Catholic believers, it will seek guidance within the sources of faith: sacred Scripture, official teaching of the Roman Catholic Church on doctrines such as that of the Eucharist or of the Shar'ia of Islam on the question of whether a pork enzyme could be used as a catalyst in a food product eaten by Muslims. Moral methodologies have also emerged in the tradition of the Church, such as probabilism and the teaching of great theologians like Thomas Aquinas. Aquinas no doubt would have been very interested in the new developments of biotechnology, were he alive today. At the same time, the final choice of whether to eat the novel food products is that of the moral actor himself or herself.

In one sense, autonomy in this matter is not really possible as long as the products of the technology are not labelled. We food consumers depend on a complex set of other human systems before we eat even the simplest item: the producers, the biochemical and seed companies that provided these producers with their inputs, the regulators in each nation who tested and approved these inputs, the retail companies who marketed the products, and the elected legislators who created the laws supporting the food industry. Under these conditions, we are not autonomous. At the same time, we would like to think that the choice is fundamentally our own, and, for some, this choice is a significant moral act.

The comment in *Nature Biotechnology* about the development of GMOs not being against the will of the "catholic God" is false in a theological sense. God is God, and is not catholic in a particularistic meaning of this word. It is also false in an ecclesiological way because the pope and his advisers have the right to comment on all moral issues, and there are many issues surrounding GMOs. Finally, it is wrong in a historical way because it was a Catholic priest-monk, Gregor Mendel (1822–84), whose basic research opened up the possibility of the science of genetics that the writers in *Nature Biotechnology* are so proud of today.

The Monk in the Garden

In her recent biography of Mendel, *The Monk in the Garden,* Robin Marantz Henig described how this obscure European monk discovered the basic laws of genetics. So obscure was Mendel and his original research that both had to be rediscovered in 1900. A depressed and introspective boy who often escaped to his bed for as long as a month to avoid the consequences of his disappointments, Mendel joined the Augustinian order near Brünn in what is now the Czech Republic. Fortunately for him and for the editors of *Nature Biotechnology*, among many others, Mendel had a gifted abbot, Cyrille Napp. Napp recognized scientific talent in Mendel and encouraged his young charge to pursue his interest in heredity.[9] At first Mendel worked with mice, but a visiting bishop found breeding rodents an inappropriate activity for the vocation of the consecrated life. Mendel turned to peas, joking: "You see, the bishop did not understand that plants also have sex."[10]

Mendel's key discovery had to do with the puzzle of why children did not always have the dominant features of their parents. He followed the practices of other plant breeders but added statistical techniques of his own to make sense of the mountain of data he had collected. He reported that features are passed from one generation to the next in peas without being blended or diluted, but that a dominant feature can mask a recessive one and that each feature's pattern of inheritance is independent of that of other features. Mendel thus distilled the secret formula of inheritance, which has become the universal ratio of genetics: 1:2:1.[11] His statistics showed that some of his yellow peas were hybrids, which meant that the recessive green feature was latent but able to reappear in later generations. He did not know why this happened, but he did know that he had made an important scientific discovery. Indeed, later generations of botanists have been able to construct reliable models for inheritance and thereby predict changes in organisms through many generations.

Mendel's work laid the groundwork for the eventual science of genetics. He did it without worry that he would transgress the will of the "catholic God," and we have all benefited from it. It is the Catholic perspective and hope that genetic modifications of plants such as Mendel's peas, whether done through breeding techniques he pioneered or through the recent methods of biotechnology, will also benefit the world community.

Issues in Biotechnology Today

Advocates of GM technology look much further back in history to find evidence of humans like Mendel who have modified plants. The plants we see in the wild evolved at the time of the dinosaurs (about 100 million years ago) for their own benefit, not for humans, and filled all the available ecological niches. For the past 10,000 years, humankind has been busy modifying a small number of these plants to serve our own needs.

Most food, even the organic variety, is the result of intense genetic modification.[12] For example, the tomato started out as a small red berry from South America that was considered toxic and used only for ornamental purposes.[13] In 1974, at the University of Manitoba, Dr. Baldur Stefansson and his colleagues transformed rapeseed, a toxic oil used as a lubricant in warships during World War II, into a health-friendly cooking oil called canola, by reducing the licosenic and erucic acids in it through conventional crossbreeding techniques.

This book is not concerned with the morality of or Catholic perspectives on artificial (humanly engineered) selection by a slow process of trait enhancement through breeding such as Mendel's work with peas. Rather, it is dealing with a phenomenon that is genuinely new: the much faster genetic modification through technological intervention, bringing a new species into the world. This novel manipulation by biotechnology is still in its infancy, and its spinoff effects are as yet unknown.

It is hoped that Catholic reflection on this question will both continue Gregor Mendel's commitment to this project and be of some help to those involved in developing these novel foods and those who are now or will be eating them.

The first chapter discusses the development of technology as a conscious choice to solve a specific problem or set of problems in a particular way. In most cases, the technique itself is neutral. What matters is how it is employed. That is to say, the use of a method can be life-giving or not, depending on the intentions of the one using it. The dual aspect of any technology is present from its start, but it becomes obvious only over time. Therefore, while we can evaluate barbed wire, for example, according to its life-giving and life-constraining properties, we cannot do this yet with any certainty for food biotechnology.

The second chapter focuses on the crucial question of "intellectual property rights" using the same "light and darkness" framework. Catholics and other Christians are clear that the right to private property is not absolute, and must give way in certain circumstances to other social values. Patents and technology-use agreements are ways that corporations make a "fair profit" on their investments in the novel foods and genetic techniques they have developed. Without the confidence that they can protect their discoveries on behalf of their stockholders, who are their owners, companies would not have made these scientific efforts in the first place. This is part of the "graced history" of biotechnology today, in which so much has been accomplished in so short a time. But patents and private property have a "sin history" as well, and these two aspects represent the "yin-yang" of GM food today.

The first two chapters offer a theological warning to readers. The third, on risks, is more positive despite the *caveat emptor* of its title. It deals with the dangers involved in the new technology and presents an argument for the ethical obligation of the purveyors of GM products to provide the consumers with clear and helpful labels. Despite widespread fear, especially in Europe, Roman Catholic thinking based on the arcane theory of probabilism in moral theology suggests to the author that we go forward cautiously in the development of new ways to deliver food and drugs to those in need of them, especially in the developing world.

The mention of probabilism might surprise some readers and raise questions of relevance in their minds. However, the Catholic moral imagination has its own manner of addressing questions like technological risk.[14] This Catholic imagination is discussed at length in the fourth and final chapter, where it is suggested that, at a subliminal level, support for the technology of gene splicing might be a déjà vu experience for some Catholics, in part because of the theological discourse around the Eucharist in which they have engaged.[15] It is further argued that the discourse used when describing the transgenic modification of wheat might provide a more useful resource for conveying Catholic doctrine on its transformation into the Body of Christ than a return to the Aristotelian–Thomistic language of the past. These historical ways of presenting Catholic teaching are difficult for many to understand today. They are puzzled by the distinction between "substance" and "accidents," which is the older way of explaining why the bread still looks like bread but is really Christ's body. But scientific discourse is basic to their thinking.

This book cannot offer a complete presentation of the topic. Much more could be said. Certain topics of great relevance, such as human cloning or stem cell research, are not mentioned. The focus is on food and pharmaceuticals delivered through GM products. No doubt other books will be written to add or to improve upon what is offered here.

Chapter 1

The "Social Mortgage" of Technology: Barbed Wire as a Case Study

Genetic modification of food uses gene splicing (recombinant DNA or transgenic technology) to solve agricultural problems. In this chapter, we will look at technology in general and this new technology in particular. It will argue that, in and of itself, transgenic technology is open to both ethical and unethical applications. Furthermore, those who profit from it do so as part of a wider community to which they are accountable; this potential liability we will call a "social mortgage." What will matter in the long run is what we do with our new technical knowledge of how to splice genes into agricultural products. Because it is so new, we will turn to the history of barbed wire as a case study for our consideration.

Most technology can be understood by a faith-informed, autonomous ethic as a value-neutral reality. Catholic theology recognizes two original sources of divine revelation, which can guide us in the use of technology. The first is formal revelation, as transmitted to us in sacred Scripture. The second is our reason guided by the faith community in which we live. It is the second source, faith-informed autonomous ethics, which applies to our evaluation of technology. Unlike the prohibition against the unjust taking of innocent human life, for example, which can be derived directly from the pages of the Bible, there is no explicit moral teaching in Scripture regarding technology.[16] From our earliest proto-human tool-making ancestors to a contemporary married couple using a digital camera to send grandparents photos of their grandchildren through the Internet, the majority in our human community have found no reasonable conflict between technological skill and their relationship with God.

But an important caveat must be added. What matters is what we choose to do with our technological expertise. Unlike nature, technology is a human creation, which we employ for good or for not-so-good purposes. A technique or skill is always used for some purpose that will have an impact on humans and other forms of life. For an observant Jew, Christian or Muslim, what we do with a particular technique, and why we do it, is of great importance.

When the gun lobby attempts to defeat licences and other restrictions on firearms, it argues that the problem is not with the technology but with the user. However, not all technologies are neutral or equivocal. Recent history has provided sad examples of technological devices intended solely for the unjust destruction of innocent human lives. For example, the Auschwitz gas chambers were developed during the Second World War for only one purpose: to kill efficiently millions of innocent humans because of their race or religious beliefs.

Clearly, both the development and the use of such technology are gravely immoral. The reason is that fundamentally a person (or community) who strives to honour God "chooses life." The author of Deuteronomy expresses it well:

See, I have set before you today life and prosperity, death and adversity. If you obey the commandments of the

22

LORD your God that I am commanding you today, by loving the LORD your God, walking in his ways, and observing his commandments, decrees, and ordinances, then you shall live and become numerous, and the LORD your God will bless you in the land that you are entering to possess. But if your heart turns away and you do not hear, but are led astray to bow down to other gods and serve them, I declare to you today that you shall perish; you shall not live long in the land that you are crossing the Jordan to enter and possess. I call heaven and earth to witness against you today that I have set before you life and death, blessings and curses. Choose life so that you and your descendants might live.... (Deuteronomy 30:15-19)

The intended use, potential benefits and foreseen harm of most technology are much more difficult to judge than the killing equipment used in Auschwitz or even the handgun. In assessing a particular technology and its applications, the difficult questions arise around what "life" means, and what choosing life with the help of this skill will mean. Usually, some applications and ramifications will clearly relate to "life," others to "death," and most to that complexity that includes both, which is the subject matter of ethics.

1. Learning from the History of Technology

In addition to the exhortation to choose life, the pages of Scripture tell us to be attentive to our communal history and to learn from our successes and failures. These precepts provide guidance in the present and for the future, and it is here that faith-informed autonomous ethics would find much light for its evaluation of gene manipulation in our food products.

Within the world of agriculture, certain patterns in the development and use of technology can be discerned. The reason for this is that the people involved, first of all farmers, and secondly the consumers of the food and other products, have been the same or similar over the course of history. Furthermore, agricultural challenges such as drought, crop and livestock pests, and the consequent need for crop protection have recurred constantly.

This book deals with a new technology. We have only recently acquired the ability to modify food grain transgenically. From what we now know, this new human skill is not related to the two "death-related" technologies of gas chambers and handguns. Instead, most parties in the current discussion seem to agree that we are dealing with a powerful discovery which, we suspect, has a dark side as well as a promise of much potential good. We have not had time to know with any certainty how "death-dealing" properties, if there are any, might emerge and how we might still choose life.

While the precise details of this ethical grey zone are unclear, it is possible to predict how the process of development may take place, based on analogies and comparisons with other related technology. One example might be the pesticide DDT. Although it is older than GMO (genetically modified organisms) technology, its history and the choices we have made about it are not yet completed, despite the fact that its use has been banned by most of the nations in the world today.[17] A better example might be "the biggest little invention of the twentieth century" that emerged from Milton, Ontario: the Robertson screw.[18] As opposed to the Phillips screw, which has a cross on its head, the Robertson variety is square. If ever there were a value neutral invention, this should be it. But one need only study the battles waged by the inventor, P.L. Robertson (1879–1951), with the Canada Screw Company over the ownership of this valuable intellectual property to realize that even a simple screw can lead to conflict on the dark side of life. P.L. Robertson filed two different patents for his invention. The first patent application was rejected by the patent office; the second one was successful, and P.L. Robertson received exclusive rights to manufacture and market his screw in 1907. A tendentious article published in 1910 in *Saturday Night*, an influential business publication, suggested that P.L. had misled investors into speculating on his Milton company when they did not enjoy patent protection for the screw. A triple-pronged defence by P.L., the Mayor of Milton and the author of an article in *The Canadian Champion* successfully laid this allegation to rest. Subsequent litigation with the Canada Screw Company upheld the validity of the second patent, and P.L Robertson is known as the inventor of the "Canadian screw."[19]

An even better example of an invention that brings out both the best and the worst in humans is barbed wire. A discussion of barbed wire within the context of the invention of new types of seeds might seem to be a category confusion, which always concerns philosophers. Genetically modified food, it would seem, belongs to nature; DDT, screws and barbed wire belong to the industrial world. However, the distinction between the natural and the mechanical within the matrix of human invention is a false one. Those involved in creating and using new seeds and barbed wire are under the influence of both divine grace and sin.

2. The Dark Side of Barbed Wire

The story of barbed wire will be presented in some detail. The reason for this is that barbed wire shares certain features with the unfinished story on which we are reflecting. These features include the following:

- both inventions were meant to address a serious agricultural problem;
- the importance of the profit motive as an impetus for development, along with controversies over ownership of the intellectual property and patent protection;

- the scale of their use;
- the global spread of the new discoveries;
- the early opposition, led by religiously motivated critics of both technologies;
- the early hints and suggestions in both of "death-related" problems, which were fulfilled in unexpected ways with barbed wire, and are feared for the GMOs;
- and, finally, the discovery of new applications for both that take their use into unforeseen areas, not at all intended by those who developed them.

Today, we take barbed wire largely for granted as a value neutral technology. Explicitly religious farmers such as the Amish and Hutterites use it without qualm. There have been no demonstrations or protest marches against it in recent memory. However, when it was originally invented in the nineteenth century, religious groups described it as "the work of the devil," and lampooned it as "the Devil's Rope"; they demanded its removal. When landowners first built fences with barbed wire to protect their crops and livestock, "fence cutters' wars" gave rise to laws being passed to ban the use of this technology in the United States. But in the meantime, there was great financial ruin and many lives were lost.

After Joseph Glidden (1813–1906) developed a successful version of barbed wire, over 530 barbed wire patents were filed and a three-year legal battle ensued. When Glidden was finally declared the winner of this contest, many companies were forced to merge their manufacturing plants or sell their patent rights to the large wire and steel companies. Readers of this book may feel they have heard this story before when they view the current struggle in Europe and elsewhere over "Frankenfood," the name given by European opponents to genetically modified organisms and novel food products.

Barbed wire illustrates the double aspect of any technology. On the one hand, it holds out hope for more abundant life, but, on the other hand, it has been associated with death as well. Barbed wire was essential to the conduct of the Boer War in South Africa, the First World War, and other recent armed conflicts. Without it, prisons and concentration camps, as we have known them in the modern age, would not have existed. The symbol of Amnesty International is a candle, representing the choice of life, wrapped around by barbed wire, representing the choice of imposing death or unjust incarceration. The reason for reflecting on a technology that most people would accept as ethical today is that a parallel history is emerging in the biotechnology field we are discussing.

2.1 Barbed Wire and Agriculture[20]

The 1870s represented a turning point in North American agriculture, not unlike the situation we find ourselves in today. The buffalo had been exterminated and, with them, the Amerindian way of life had also been extinguished. A new order emerged in the West.

This change was not immediately evident in the United States because the Texas longhorns, descended from cows brought to America by the Spaniards and allowed to run wild, replaced the buffalo. These animals were similar in many ways to the buffalo they replaced. The so-called cowboys replaced the Amerindians, and moved about on horseback, herding these cows annually and driving them east to be killed.

Grass had provided the fundamental source of energy for the buffalo, and also for the cattle, which numbered more than 11 million by 1880. On the surface, this transformation from buffalo/Amerindians to cattle/cowboys might suggest merely one of species and ethnicity. The truth is that a new era of animal husbandry and agriculture had begun, and with it a new problem, requiring a technological solution. The problem was how to farm the land while, at the same time, manage these large herds of animals. What made cows so valuable was that they could find their own pasture. The problem was how to restrict their freedom of movement while retaining their profitability.

Initially, the West was not fenced at all. Fences were not needed, because these cattle were self-sufficient and were allowed to run freely until the round-up and transportation east for slaughter. A small number of men on horseback could collect them and take them to market. Their economic attraction was that they did not need constant supervision. Ranchers controlled their cattle by geography. They assigned grazing areas along riverbanks. Branding them identified ownership.

In the late nineteenth century, all of this changed. Both the U.S. and Canadian governments offered incentives to farmers who wanted to settle on the land. In the U.S., the Homestead Act of 1862 offered 160 acres to private settlers after they had lived on the land for five years. Similar legislation was passed in Canada. The communal choice to expand agriculture westward created the need for capital investment in fencing. In 1871, the annual expenditure in the U.S. was estimated to be $200 million, and a U.S. Department of Agriculture report stated that nearly $2 billion had already been spent on fences. Over the decades, for every dollar invested in livestock, another dollar was required to protect crops from wandering animals.

Fencing materials presented a problem. Possible materials included wood, stone and Osage orange hedges. But none of these options was straightforward. Wood was rare; the right type of stone wasn't available; and the hedges were difficult to transplant, and took three or four years to grow to their full height. The developing railways provided a stopgap solution insofar as wood was already being shipped from the East for house building, and the railways themselves needed it for bridges and other purposes. However, because of the vast spaces that needed fencing and the cost and inefficiency of shipping, this solution was ultimately unprofitable. A new type of fencing was urgently required.

The first attempt at a technological fix seemed to be a failure, but such false directions often lead to the solution. In 1873, Henry Rose, who had a farm in Waterman Station in Illinois, came up with a new way to control a cow. He attached a wooden board equipped with prickly pieces of wire to the cow's head, so that if the animal were to squeeze through a narrow gap in the fence, it would cause itself pain. It later occurred to Rose that the wire could be attached to pieces of wood along the fence itself. The experiment worked, and the cow learned not to go near the fence. Other farmers conducted similar trials, supplementing wood with iron barbs. We remember the name of Henry Rose because he took the trouble to patent his idea and to display it at a farm exhibition in De Kalb, Illinois.

By 1874, six different barbed wire patents had been granted, and by 1880 something like 50,000 miles of barbed wire fences kept cattle in their place. Marketing the idea was not at all difficult and involved a flair for the dramatic. At a county fair in San Antonio in 1876, an entrepreneur sealed off part of the town's central square with barbed wire fencing and filled it with dangerous-looking longhorn steers. When the animals charged the barrier, the pain of the metal tearing into their skin infuriated them. They attacked the fence with renewed vigour, only to suffer further wounds. Eventually they backed away, having learned that this new technology could not be overcome. In a remarkably short time, there were few farmers in the American or Canadian West who did not know that barbed wire offered a cheap, adaptable and sure way to protect their valuable crops from cattle, without human supervision.

Barbed wire could stop animals, no matter how many they were, or how desperate to roam beyond the fence. At first, manufacturers, anxious to guarantee the outcome, produced large and sharp barbs. When cattle crushed against the wire, as they did in San Antonio, they were seriously injured and in warm, humid climates, open wounds led to screw-worm infections. This was especially troublesome to the range cattlemen who did not benefit from barbed wire themselves, but saw their stock harmed by its use. Farmers who preferred to control their animals without barbed wire also suffered when wire was erected by the railroad, for example. Legislation to prohibit its use was introduced in several states, but was always defeated in the western states, although it became law in some eastern states. The compromise was that gradually the barbs became less sharp, or vicious, to use the technical term.

As humans became more sensitive to animal pain, so animals learned more about the propensity of humans for violence. A commercial leaflet encouraging the use of barbed wire recommended that the farmer lead young horses "to the fence and let them prick their noses by contact with it...they will let it thoroughly alone thereafter." Farmers seemed concerned to caution their horses about barbed wire, but cows were expected to learn about it through hard experience. The fact that animals seemed able to learn from their experience led manufacturers to produce more conspicuous barbs, which they marketed as more obvious; the new design functioned as an indirect form of intimidation.

By the end of 1880, the success of the technology seemed complete, and it had spread throughout the world. Washburn and Moen, the leading American manufacturer of barbed wire, could claim that they had established patents covering Australia, New Zealand, India, Italy, Sweden, Austria and Denmark, "5,470,952 miles in which no barbed wire can be sold without direct infringement.... A territory compared with the territory of the United States as two is to one," and they expected "300,000 tons of sales per annum."

By 1914, barbed wire had become a common feature of global agriculture. In Europe it protected crops from animals, and animals from other animals. Elsewhere it helped to transform land that was not yet firmly under human control. In Australia, for example, barbed wire was used to exterminate unwanted wild animals simply by fencing off their sources of water. With the spread of the railways, barbed wire truly became a global technology because it was used to protect the track from stray animals. From 1899, the leading manufacturer was the American Steel and Wire Company. In its first eight years, 34 per cent of its production was exported; in the next eight years, 44 per cent was sold abroad. But the American market, where more than 100,000 tons was consumed annually, was still the mainstay of the demand.

And then an entirely novel use for the technology was discovered, in which its use became even more directly related to power and control, this time not by farmers and ranchers over animals, but by one group of humans over another.

2.2 Barbed Wire and War

In October 1899, the Boers of the Transvaal and the Orange Free State declared war on Britain. The Boers (the Dutch word for "farmers") were early settlers in South Africa who were of Dutch or Hugenot descent. The South African War, often called the Boer War, should not have lasted long. By June 1900, the important Boer towns had been captured, and the situation seemed to have been normalized. It had been a conventional war, with fortresses besieged, soldiers deployed and battles won or lost. But this war was not yet over, and what followed was in no way conventional.

Falling back to regroup, the Boers concocted a new strategy, which today we call guerrilla warfare. Small, highly mobile units of soldiers on horseback, known as commandos, began to harass the British, cut the rail lines and stage small-scale ambushes. The purpose of these tactics was twofold: to make the invasion of their homeland as costly as possible for the British, and to encourage the Boer communities of the Cape and Natal to join in the rebellion. The British could not occupy South Africa, they reasoned, if the whole country fought against them.

For the British troops, winning the war meant containing and overcoming the last of the Boer soldiers, and preventing the guerrillas from entering British-controlled territory. They no longer needed to win battles or mount sieges. However, ending the conflict meant enforcing an armed occupation. To prevent the local population from offering aid and comfort to the Boer commandos, who were mainly farmers, civil society was broken up. Farms were destroyed and civilians were gathered into what were known as refugee camps or concentration camps.

The British were mobile, thanks to the railway, but because of their horses the Boers were mobile too. The side that could move more quickly and effectively would win. In order to protect their supply lines and to block the movement of horses, the British fortified the railways with a complicated pattern of parallel barbed wire fences, stretching thick entanglements between them. Although humans could cut through, the barbed wire prevailed against horses. It took time to cut the barbed wire, and the guerrillas who tried to do so presented stationary targets.

The British constructed blockhouses, small forts about a mile apart. They surrounded them with thick barbed wire entanglements, too, and stationed six or fewer soldiers in them. The soldiers had a new weapon – the machine gun. To cross the lines, the Boers had to make their approach as far away as possible from the block houses and cut the wire as quickly as possible, all the while under machine-gun fire.

By 1902, the veldt, always sparsely populated, was now almost empty. It was criss-crossed by 35 barbed wire lines, next to the railroads and going beyond them – the longest being 175 miles in length. There were about eight thousand forts and thousands of miles of wire.

A few years later, Japan imitated the use of barbed wire by the British in the Boer War when it invaded Manchuria in February 1904, and in the war with Russia that followed. Both Russia and Japan had to transport their soldiers great distances across barren land in a harsh climate. Although the troops were equipped with modern weapons, they were slow to reach their destinations. The war was characterized by unhurried buildups combined with intense and rapid attacks on a well-entrenched and fortified enemy.

Trench warfare really began with the Russo-Japanese War. Seven to eight feet deep and lunette-shaped, the trenches held, on average, one man per yard. Although they were not continuous, as they later became in World War I, they represented a crucial line of defence. A thick barbed wire entanglement was set up about sixty yards in front of the forward trenches, which contained machine guns. These guns were effective at a range of a few hundred yards and would prevent attacking forces from advancing even if there had been no barbed wire. Soldiers stuck in front of the entanglements were slaughtered. The price paid in lives for cutting through the entanglements was great. Barbed wire was even laid inside trenches. When Japanese soldiers managed to cross into a Russian trench, they found that the retreating soldiers had thrown "cages" of barbed wire down behind them to prevent further advance. The land war ended in a bloody draw which neither side won. In the end, this war was won at sea by the Japanese navy.

In the South African War, barbed wire played an important strategic role by protecting lines of communication and large areas of land. In the Russo-Japanese War, it played a decisive tactical role by creating small areas that could be defended with machine-gun fire. In World War I, the strategic and tactical functions of barbed wire were united. Following a rapid advance, the Germans were halted in September 1914 along a line of defence that stretched across northeast France. The trenches that both sides dug resembled those of the Russo-Japanese War except that they used even more barbed wire. The Germans depended on barbed wire even more than the Allies. Reviel Netz quotes J. Ellis as saying, "Their wire was hardly ever less than fifty feet deep, and in many places it was a hundred feet or more. In the Siegfried Line every trench had at least ten belts of wire in front of it." A belt was a single entanglement consisting of numerous strands of wire.

The warfare advantage of attack and mobility was restored by the new technology of the tank that rendered the barbed wire of World War I outdated.

2.3 Barbed Wire and Human Populations

In 1896 during the war in Cuba, Spanish General Valeriano Weyler ordered the entire population of the island to congregate inside fortified towns. He was fighting a guerrilla army, and this strategy was the only way he could think of to cut off the guerillas from their supporters. Hundreds of thousands of people were forcibly relocated. The apparent brutality of this strategy played a part in preparing the American public for the Spanish-American War of 1898, when most of Spain's overseas possessions were taken over by the Americans. The following year, the Philippines erupted in revolt against U.S. domination. The American strategy was "reconcentration." Barbed wire was not used in these military actions, but the tactics adopted presaged developments in the twentieth century.

Meanwhile, as the South African War continued, more and more Boer farms were destroyed and their inhabitants removed. There were no cities or villages to which they could be relocated, so the British created a whole apparatus to provide them with accommodation, medical care, education, protection and food. It consisted of small, temporary settlements known as concentration camps.

In all, there were about fifty camps, each holding a few thousand inmates. Almost all were women and children; the men were out fighting the British. Deaths from disease reached catastrophic numbers, estimated at thirty thousand overall. The camps consisted of family tents, with a few larger tents containing kitchens, toilets and hospital beds. Because there were so many children and the inmates were dependent on the authorities for everything, control was not a problem.

At first, the camps were left unfenced because the British wanted to pretend that the inmates were refugees who had come to them for protection. Kendal Franks, a British inspector, reported that when the Boers raided the camp at Standerton in the Transvaal on August 11, 1901, and captured 157 cattle, the camp supervisor, Frank Winfield, ordered the inmates to erect a fence. This was the first recorded human settlement whose boundaries were defined by barbed wire. Several weeks later, Franks visited the camp at Volksrust and noted that its five thousand inmates were surrounded by a double perimeter fence of barbed wire, which rendered a local police presence superfluous. A group known as the Committee of Ladies, appointed by the British Government to inspect the camps, actually recommended that fencing become compulsory because it made cleanliness within the camp easier and prevented infectious diseases from reaching the general public. British propaganda always stressed the unhygienic practices of the Boers, and explained that superior British methods were to be imposed for the Boers' own good. What is worth stressing is the ease with which the fences were erected. A medium-sized settlement could be surrounded by wire in a few days. The material was readily available; the investment in labour was not excessive. In a certain sense, barbed wire camps were created because barbed wire was available.

In World War I, barbed wire became a standard feature in prisoner-of-war camps. The prisoners could be rapidly moved, often by train, from a temporary pen to other, permanent enclosures surrounded by a barbed wire fence about ten feet high. Sometimes the camps had an extra fence fifty to seventy-five feet from the first. Prisoners found between the first and second fences would usually be killed. The guards who were posted in towers around the circumference found the partial transparency of the barbed wire useful.

During World War I, barbed wire was used to control citizens as well as captured soldiers. About thirty thousand British residents held German citizenship, and it seemed obvious to all that they should be interned. Some were placed in unused buildings, such as former factories or empty skating rinks. But most were simply put behind barbed wire in two camps on the Isle of Man. The concentration camp had become the standard way to control enemy populations, civilians as well as soldiers.

In a civil war, an enemy is not defined by nationality. In the USSR, more than twenty-four thousand individuals perceived as agitators against the Soviet State were detained in 56 camps by 1922. After the Civil War ended and the Bolsheviks were fully in charge, greater use was made of existing structures. For example, the Solvetzsky Islands in the White Sea, with their monastery-like buildings, were, for a time, the main incarceration centre in Russia. The number of prisoners on Solvetzsky increased from 3,000 in 1923 to 50,000 in 1930.

At the end of the 1920s, Russia began to prosecute a war against the "Kulaks" and "saboteurs." In 1933, there were 2.5 million people in the *gulag*, the generic name given to the prison camp system in Russia, and this number doubled by the time Stalin died in 1953. Although the camps never again regained their prominence, they remained part of the Soviet landscape, and no one knows how many of them were built. The gulag's perimeter, if stretched from end to end, amounted to tens of thousands of kilometres, a figure comparable to the length of the railroads and barbed wire fencing in Canada at the beginning of the twentieth century. The layout of the camps was standardized: layers of barbed wire fences, punctuated by guard towers.

If barbed wire has acquired a thoroughly vicious reputation as a way of controlling human populations, it is not because of what the British did to the Boers, nor the Bolsheviks to the "saboteurs." This technological solution to the problem of keeping perceived enemies in a secure place is abhorred primarily because of the Holocaust.

The history of German concentration camps began in 1933. Like the class enemies of the Bolsheviks, the Jews were seen as enemies who had to be controlled. Their movement was

restricted initially by legal means. The comprehensive codifi-
cation of the Nuremberg Laws of 1935 barred them from certain economic activities and from contact with Gentiles. Some
suggested resettling Jews in some clearly defined area such as
the Island of Madagascar.

The eventual plan was much more extreme. After Nazi vic-
tories in Poland, ghettos were set up, comparable to the concentration zones of Cuba and the Philippines but intended exclusively for a particular human group and defined on the
ground by barbed wire. By 1941 Jews were being systemati-
cally murdered in the occupied areas of the Soviet Union, and
at the end of that year, the Final Solution culminated in its
most notorious phase, the death camps.

Within the context of total horror, barbed wire might seem
to be an insignificant detail. In fact, it was the central element
in the architecture of the death camps. "We're separated from
each other by barbed wire," Wolfgang Sofsky says, quoting a
survivor in *The Order of Terror*. "The Germans evidently have a
very special affection for barbed wire. Wherever you go, barbed
wire. But you gotta give credit where credit's due: the quality's
good, stainless steel. Densely covered with long barbs. Barbed
wire horizontally, barbed wire vertically."

Barbed wire was used not only on the perimeter but also
within the camps themselves. At the perimeter was the cus-
tomary multi-layered fence. However, the transparency of wire
began to seem a disadvantage because of the need to hide what
was going on inside, so concrete walls were added.

But the prisoners felt that what surrounded them and con-
fined them above all was not walls but wire, for two reasons.
First, the wire was electrified, making it a doubly horrible symbol of the violence engulfing them. Second, conforming to a
version of the rules governing prisoners of war, an inmate who
went within a certain distance of the barbed wire would be
shot immediately. Reviel Netz argues that the ultimate purpose
of barbed wire in the death camps was to extinguish any

private space at all, so that the presence of the Jews disappeared even before their actual murder.

2.4 Barbed Wire as a Human Choice

The history of barbed wire did not end with the death camps. It continues to have a multitude of uses, and the industry continues to be a dynamic one. The literature on it takes two directions. One is concerned with its human context, because barbed wire is a symbol of oppression. The second direction discusses it in technical detail, and focuses on its use to control animals. In this context, barbed wire is a symbol of strength and ingenuity.

In fact, the extension of barbed wire from the control of animals to the control of human movement was a clear choice and an obvious development of the capacity of this technology. At its ultimate reality and meaning, where barbs tear at flesh, there is not a great deal of difference between humans and animals entangled in wire. The erection of barbed wire in the Canadian West did not lead in a direct way to its use in trench warfare nor to the concentration camps, but the technology itself has a potentiality which has led down that road. Osage orange fences could not have surrounded Auschwitz effectively. A technology was needed, and it is not clear that it would have been barbed wire, were it not for the invention of Henry Rose and the patent of Joseph Glidden.

At the time of the American Civil War, the materials required to make barbed wire were all available. That war spawned all kinds of military technology, such as machine guns and submarines. Furthermore, the conflict involved large numbers of soldiers, and control of POWs was a major challenge. The challenge culminated in the notorious open-field prison of Andersonville. In fact wire, non-barbed wire, was used as a tactic to create obstacles in a field. But that use remained local, was subject to availability, and differed fundamentally from the subsequent use of wire with barbs on it. Barbed wire entered human history as a method of exercising power and control of one group of humans over another in the South African War, and not before then. A choice was made to create a technology and then to extend it to this use. For good or ill, barbed wire may always be with us.

3. Catholic Teaching on the "Social Mortgage"

Initially, the invention of barbed wire solved the agricultural problem of combining farming with the existing ranching on the North American plains. In a similar way, GMOs are intended for crop protection and enhancement in order to solve the problem of infestations of various types. Both technologies are intended to improve our management of the process of food production. The inventors of these technologies and their various applications rightly consider these ideas to be their intellectual property and have taken out patents on them. The owners then sell or license the use of their inventions.

This also is legitimate, as long as certain caveats are respected. It is characteristic of Christian social doctrine that the goods of the world are originally meant for all. As Pope John Paul II has put it, "The right to private property is *valid* and *necessary*.... Private property, in fact, is under a 'social mortgage,' which means that it has an intrinsically social function, based upon and justified precisely by the principle of the universal destination of goods."[21]

Humans seek ownership of the instruments of control over *external* reality. The notion of "social mortgage" suggests that certain problematic applications of the original inventions, particularly those involved in control of the *internal* reality of human beings, especially with regard to their health, nutrition and, above all, their freedom, are the responsibility of the owners of the technology, their successors and those who have profited from them. The reason is that modern technology consists in a complex series of creative breakthroughs, applications and further discoveries. Ownership of the process is shared within a technological community, but responsibility for unintended results is also shared.

Ursula Franklin divides technologies into two types, according to the process that they involve: holistic and prescriptive. Holistic technologies are often related to the crafts created by artisans who control the process of their work from beginning to end. It is their hands and minds that make decisions such as the thickness of the pot or the height of the doorway into their house. The opposite type of technology is specialization by process. Here the making or doing is broken down into clearly identifiable steps. A different worker or group of workers, who need to be familiar with the skills of just that one step, accomplish each of these steps.[22]

Both barbed wire and GMOs are prescriptive in Franklin's sense of the word: they have creatively focused the human endeavour to produce more and better food with less loss in a piecemeal or step-by-step way. Henry Rose's invention and patent gave way to that of Joseph Glidden. The use of barbed wire by the railways to protect their tracks from livestock was not on their minds. Similarly, when barbed wire began its second application as an essential instrument for *internal control* over the movement of human beings and their self-image as prisoners or slaves, Rose, Glidden, and even Washburn and Moen, the American manufacturers of barbed wire, had not made this ethical choice. If in some way they could have seen a vision of the concentration camp at Dachau they would, no doubt, have been deeply troubled by that nightmare application of their invention.

With its use to control human beings, the technology of barbed wire had crossed an ethical line, one that has not (yet) been approached by the GMOs. Due to barbed wire's nature as a prescriptive technology, none of the inventors nor those involved with its initial applications could stand accused of choosing "death" rather than "life." Despite this, by the beginning of the twentieth century, the barbed wire project had changed and was open to even more ethical objections than at the time of its invention, when it was called the "Devil's Rope." When the technology began to be applied to the control and intimidation of humans, this opposition no longer existed, perhaps because barbed wire was a fact of life. It is possible that the same dynamic may happen with GMOs.

This suggests that the responsible use of prescriptive technology requires us to clearly recognize that the necessary ethical criteria cannot be provided by the technology itself. When Henry Rose and Joseph Glidden made their discoveries, the later human uses of their ideas could not have been articulated if we merely asked *what* barbed wire does. However, if we were to ask *how* it accomplished its intended purpose, we would have had cause for reflection even at the initial stages of its development.

Both barbed wire and the GMOs are exceedingly effective and efficient. They produce predictable results. The technology of barbed wire itself became an agent of ordering and structuring, not only of animals, but eventually of humans as well. This has not (yet) happened with GMOs. During and after the Boer War, barbed wire helped to create a culture of compliance and conformity among those incarcerated in the camps. People were programmed.

In other words, like it or not, the development of barbed wire entailed what Pope John Paul II has described, within other social contexts, a "social mortgage." And there probably will be a similar one that will have to be paid for the development of techniques to accomplish the transgenic modification of food.

Ethical criteria are open to dispute. The possible infliction of pain on animals and on humans that is part and parcel of the process of laying down strands of barbed wire is open to different evaluations. But with prescriptive technologies, serious questions need to be asked. As a matter of fact, these questions were asked with regard to barbed wire and are being asked now about the transgenic modification of food. Unlike DDT, neither has been banned, nor will they be. They have proven uniquely able to solve serious agricultural problems. However, they do bring attendant social and environmental problems that will have to be solved. The cost of these solutions must be estimated and then paid as our "social mortgage" of technology. Those who live with actual mortgages on their own real estate might wonder who pays this social mortgage, when and how. They need to realize that the "analogical imagination" of Catholics, which we will discuss in the last chapter, cannot answer such questions.

Chapter 2

Intellectual Property Rights in Technology

Much more could be said about the moral history of barbed wire. But only one further connection will be made between its story and the emerging one of transgenic modification of food. This connection has to do with patents and the more general topic of intellectual property rights.

It is generally acknowledged that the "Father of Barbed Wire" is Joseph Glidden.[23] But that simple statement, though true, summarizes one of the most controversial chapters in the history of patents in the United States. Before Michael Kelly "twisted two wires together to form a cable for barbs – the first of its kind in America,"[24] there had existed fences made of one strand of wire, but they were easily broken by the weight of cattle pressing against them. Had Kelly properly patented his invention he, not Glidden, would have been the "father." He did not.

Ten years later, in 1873, Henry Rose of De Kalb, Illinois, improved upon Kelly's core idea by introducing a new concept in fencing, a wooden rail with a series of sharp, protruding spikes that could prick an animal that came too close.

Three people saw Rose's invention and were impressed by it: Joseph Glidden, Jacob Haish and Isaac Ellwood. Ellwood was unsuccessful in his effort to improve upon it. Joseph Glidden, more successful, filed for a patent in late 1873, but his application was rejected. Later, in 1874, Jacob Haish patented a form of wire known as the "S-barb." However he did not make a serious effort to promote and sell it. Meanwhile Glidden had established the Barb Fence Company to produce and market his version of barbed wire. Haish sued Glidden. In the subsequent legal battle the courts decided for Glidden. They reasoned that Glidden's patent had been filed earlier than the one by Haish, and ruled in its favour because it both described the method for locking the barbs in place between the two twisted wires and the machinery to mass-produce the wire. Therefore his 1873 patent application should be accepted. In the end, Glidden prevailed over Haish and went on to make his fortune. Even though the legality of Glidden's claim was upheld in the courts, the ethics of his ascendancy over Kelly, Rose and Haish remain murky to this day. It is therefore to this complicated but crucial question of intellectual property rights that we now turn our attention.

If Joseph Glidden is credited with the invention of barbed wire with techniques for its mass-production, and if this knowledge was his own intellectual property, we must consider a similar case for GMOs. The new technology of genetic modification of organisms is associated with the 1953 description by James Watson and Francis Crick, both scientists working at Cambridge University, of the molecule that contains genetic information, known to us as DNA. Crick and Watson correctly stated that DNA takes the shape of a double helix. The equivalent of Kelly, Rose and Haish in the story of this discovery would be one woman, Rosalind Franklin (1920–58) of King's College in London. In 1951, Watson attended a lecture by Franklin on her work to date on DNA. Using X-ray diffraction images of DNA and working mostly alone, Franklin showed that in higher humidity DNA had the characteristics of a helix. She suspected that all DNA did so as well, but she did not want to announce her discovery until she had evidence for the shape of DNA at lower humidity as well. This frustrated her supervisor, Maurice Wilkins, who showed Franklin's results to Watson in 1953, apparently without her knowledge or consent. Crick later admitted, "I'm afraid we always used to adopt – let's say, a patronizing attitude towards her."[25]

Watson and Crick made a further conceptual step, based on Franklin's work, and suggested that the molecule consisted of not one but two chains of nucleotides, each a helix, but one going up and the other going down. Also, each strand of the DNA molecule was a template of the other, and it was this discovery, above all, that has led to the revolution in biotechnology that we are considering in this book.

For a variety of reasons, Franklin left King's College to head her own research group at Bilbeck College in London, where she worked not on DNA but on coal. By 1962, when Watson, Crick and Wilkins won the Nobel Prize for physiology/medicine, Franklin had died of cancer. The Nobel Prize goes only to living recipients and can only be shared among three winners. Were she alive at that time, would she have been awarded the highest scientific honour in place of Wilkins?[26]

1. Patents and Profits

Within the world of intellectual property rights, one discovers a battle between grace and freedom, on the one hand, and sin and deception, on the other. For this reason, any discussion of Catholic thinking on justice in relation to the current questions surrounding biotechnology must address itself to intellectual property rights. By property is meant not only material objects but also thinking, such as the creative ideas that lead to new ways of producing the novel foodstuffs that have been transgenically modified. Both newly invented things and ideas or new methods of doing things are protected legally by patents for a certain length of time in most countries. International agreements and regulatory bodies try to ensure that patents in one country are respected in other countries. For various reasons, intellectual property rights and patents are significant issues.

Monsanto, Aventis, AstraZeneca (Sygenta), Dupont, Novartis and the other multi-national corporations responsible for much of the current development in the area of biotechnology in agriculture and pharmaceuticals insist that the involvement of their research scientists and supporting resources depends upon assurances that their owners, the stockholders of their companies, will recover the heavy expenses involved in research and development of genetically modified food, as well as make a profit as good or better than they could from other available investment options.[27] Crucial to this recovery of costs and the production of new wealth for their owners is the guarantee that their intellectual property, in the form of the novel forms of food, is protected from competitors, and that those who use these products or procedures pay for them. In other words, these managers and executives need to turn a fair profit for those who have chosen to invest in their company rather than in another wealth-generating enterprise – hence the importance of patents.

Much depends on the meaning of "turn a fair profit." Many would argue that food is the staff of life. The new possibilities for growing more food in poorer agricultural conditions mean the difference between life and death by hunger, disease or starvation for millions. Also, "biopharming," or the production of drugs in genetically altered plants, offers new hope to disease-burdened countries in Africa and elsewhere. Therefore, they argue, this technology should not be included under the existing legal framework of patent protections. Put simply, farmers in places that need to put GMO technology to work may not be able to afford the seeds and other proprietary products, especially if they cannot reuse the seeds saved from the first crop for future ones.

The subsistence farmer who is also a seed saver is not the only person who has problems with the way intellectual property is protected today. Many are troubled by the very idea of patenting life forms. On June 16, 1980, the Supreme Court of the United States ruled in favour of a civil suit initiated by Amanda Chakrabarty, a microbiologist at the University of Illinois Medical Centre, who had invented a strain of GM bacteria designed to digest crude oil. Dr. Chakrabarty's discovery would have obvious environmental importance, but Sidney Diamond, the U.S. Commissioner of Patents and Trademarks, refused to grant her a patent on the grounds that, under existing U.S. Patent Law, "living things" were not patentable. In its *Diamond* decision, the U.S. Supreme Court disagreed.

A similar struggle has gone on in Canada. There, plant breeders had been happily modifying the way things grow long before the recombinant DNA revolution in agriculture. Their intellectual property was not protected under existing laws. The legislation regarding patents restricted the items that were subject to protection in the following ways:

1. The creation must be an "art, process, machine, manufacture, or composition of matter";
2. It must be of human invention and not a product of nature;
3. The creation must be new, useful, and non-obvious.

In other words, until the recent "oncomouse" litigation, Canadian courts had not accepted that plants and animals were patentable.[28]

Canadian patent law held out against patenting life forms until August 2000, when the Federal Court of Appeal overruled Canada's patent commissioner, as well as its own trial division, and ordered that a patent should be issued on "the genetically engineered, cancer-prone creature that has become known as the Harvard Mouse – in fact, on any 'transgenic non-human mammal' whose cells carry the modification that makes it useful in cancer research."[29] Lawyers for the Federal Government of Canada have appealed this decision to the Supreme Court.

At the time, only a few people recognized the implications of the *Diamond* decision. One was Jeremy Rifkin, then the Director of the People's Business Commission in Washington, D.C., who predicted that "In 10 or 20 years, *Diamond* will be looked upon as one of the biggest decisions a court has ever made." He added that holders of a patent on gene-splitting techniques "would be able to play God."[30]

Diamond has proven to be as significant as Rifkin predicted. Patents continue to be a battleground. For example, in December 1999, an obscure bureaucrat in the European Patent Office mistakenly issued the first patent to allow the cloning of a human being. Even more troubling, say critics, is the fact that this patent may have given scientists the licence to change the genetic makeup of the entire human species. "The patent, issued in Munich, No. EP-0695351B1, gives its holder, the University of Edinburgh in Scotland, the sole right to make, use, and sell human beings created in its laboratory."[31] One way to understand patents is as "rewards from the state for sharing a new invention with the public, so the disclosure must allow others access to the invention."[32] If this is so, then the University of Edinburgh, where Darwin once studied, simply wanted to patent a set of laboratory methods that, in time, might prove useful in fighting diseases like Parkinson's. The European patent commissioner intended none of this either, but during the six years of patent review that EP-0695351B1 underwent, someone forgot to insist on the necessary legal disclaimers being added. Greenpeace, the environmental group, discovered the botched patent and mounted a noisy demonstration that resulted in the patent office acknowledging its error.

Since 1986, the annual number of U.S. patents granted for a gene or gene fragment has increased by more than 6,519 per cent.[33] Some patents may be against the perceived common good. Should "private companies be given exclusive rights to exploit for their profit the codes printed by nature in the cells of every human being"?[34] For instance, Myriad Genetics, an American company located in Salt Lake City, has patented two human genes crucial for the screening of breast cancer, BRCA1 and BRCA2. Much of the work on the second of these genes was done in Britain at the Sanger Centre in Cambridge and the Institute of Cancer Research (ICR). Myriad filed its patent application just hours before ICR published its results on BRCA2 in the journal *Nature*, and ICR continues to insist that it discovered the gene first.[35]

Within the Canadian agricultural context, the struggle has been to strike a suitable balance between the scope of ownership rights of those who claim possession of a newly developed plant variety and the residual freedom of farmers and breeders to use this creation for their own purposes. In 1990, Canada enacted the Plant Breeders' Rights Act, a patent-like piece of legislation that is part of a network of such rights around the world.

Critics of this legislation complain that it concentrates ownership of these "new" seeds in the hands of a few. Furthermore, the Canadian legislation gives no recognition to Third World countries that are the genetic storehouses upon which plant breeders depend. Finally, such a law might hasten and encourage genetic uniformity and erosion.

Germplasm, the genetic material used by breeders, is thought by these critics to be part of the common heritage of humankind and should be available without restriction. Legislation such as Plant Breeders' Rights, it was feared, would result in a reduced number of crops, and place our food supply under threat of drought, disease and insect infestation.

Even before the GMO revolution of the last fifteen years and the development of legislation in response to it, such risks were realized. In 1970, half the orange crop of southern Florida and Texas was destroyed when a virulent new fungus attacked it. Ireland's famous potato famine of 1845 was the result of a Mexican fungus that destroyed the genetically similar potatoes. It is also thought that the ancient Mayan civilization of 900 CE was hit with crop destruction due to the genetic uniformity of its maize crop.[36] Examples of this sort could be multiplied.

The simple answer to these concerns is that the Plant Breeders' Rights Act is the "least worst" solution at the present to many ownership issues, and that genetic erosion is not someing new. Because we are focusing here on the relationship between traditional Catholic teaching and novel GMO foods, we look at the issues of seed piracy, the "terminator gene," and other technology protection systems such as genetic use restriction technologies (GURTs), and the employment of technology use agreements (TUAs) by Monsanto and other seed companies. To provide an analytical horizon for future, more socially responsible directions that the use of patents may take, we will explore some developments in the pharmaceutical industry. We will examine particularly the decision of Monsanto to offer a patent waiver on "golden rice," a new variety of genetically altered grain that contains extra beta carotene, a molecule that helps the body produce vitamin A and thereby prevent blindness. This move will be a great help to malnourished children who suffer from vitamin A deficiency.[37] The author's view is that the possibility of molecular farming that will deliver both food and necessary drugs to those in need provides us with hope that this technology will prove to be of lasting benefit to our world.

2. The New Sins: Seed Piracy and "Terminator"

Those involved in the global effort to defend and develop the rights of women refer from time to time to the sin of sexism. By this they mean the deliberate decision to discriminate against women, for example, by restricting their career choices and job opportunities. Among other things, their opponents might ask them where this new, never-before-heard-of "sin" originated. Could a new sin appear in the late twentieth century, of which we were never previously aware?

There are at least two answers to this problem. The first is that cultural ethics and the applications of traditional church teaching are not fixed and final, but develop in such a way that new problems and "sins" can emerge to engage our communal conscience. The second is to point out that the sin of sexism has indeed been discussed in the past under other names. Literally millions of people feel the pain of sexism; relatively few are aware of the legal and ethical issue described by the world's second-largest seed company, Monsanto, not as a "sin" but as "seed piracy." Like sexism, this form of injustice is a new variation on an old theme. In medieval theology, the malice of theft was described graphically as follows: "*Res clamat domino* – the [stolen] thing cries out to its owner." According to Monsanto, many genetically modified, patented seeds are doing the same, and they, the board, employees and stockholders of their company, are the owners of these seeds and their produce. In what follows, the moral issue of seed piracy is shown to be an example of the general problem of intellectual property rights in the world of biotechnology.

Modern agriculture increasingly encourages monoculture. This means that a farmer will plant one crop exclusively. This planting method and the use of biotech crops are associated with the problem of susceptibility to weeds and the consequent need to spray ever greater amounts of herbicides. In the 1980s, Monsanto, the world's third-largest agrochemical company, made a great deal of money from the popular herbicide Roundup™, whose active ingredient is glyphosate. This herbicide attacks a crucial enzyme in plants that is not present in animals and humans. Hence the poison was not considered harmful to them. Glyphosate is very powerful but is not selective. It kills all the plants it touches. As the legal lifetime of Monsanto's patent on glyphosate approached, many scientists, including Monsanto researchers, began trying to find plants with natural Roundup™ resistance or to develop mutant plants in order to create a new application for the herbicide and a new genetic possibility. Using cells from the petunia, scientists at Monsanto were able to isolate and clone an altered gene that was less recognized by glyphosate than other genes that it killed, but was still functional in making the amino acids that give the petunia its distinctive aroma. Monsanto then had the first "herbicide-tolerant gene" that was immune to Roundup™.[38] By inserting this gene into its own seeds, the company made it possible for farmers to spray herbicide on a field planted with its genetically altered seeds and kill all the weeds but leave the crop itself untouched. This was thought to be a good way to make a profit on both the seeds and the herbicide. Such, indeed, has proven to be the case. Roundup™ and Roundup Ready® seeds account for at least 20 per cent of Monsanto's annual sales.[39] Although the seeds can cost more than U.S. $32 per bag, farmers seemed ready to pay this amount in the hope of higher yields and of savings on expensive inputs of herbicides and fertilizer. All of this depends on a social and legal contract between the company and the farmers, known as a Technology Use Agreement, that precludes saving and reusing the seeds another year.

Four years ago, 30,000 farmers who used Monsanto's Roundup Ready® soybeans allegedly received a letter warning them not to save and replant seeds from genetically engineered crops and describing this practice as "piracy." The letter reads as follows:

You may have heard about recent investigations in your area concerning farmers, or others helping farmers, illegally saving and replanting patented biotech seed such as Roundup Ready® soybeans. Saving and replanting seed with these patented biotech traits is seed piracy. Throughout the 1998 growing seasons many growers, retailers, seed conditioners and others have communicated their interest in learning what Monsanto is doing regarding seed piracy prevention.

Following a recent seed piracy investigation, David Chancy of Reed, Ky., admitted to illegally saving and replanting Roundup Ready® soybeans. Chancy also acknowledged that in return for other goods, he illegally traded the patented seed with neighbours and an area seed cleaner for the purpose of replanting. All of those involved were implicated when Monsanto made the discovery.

Chancy's settlement agreement terms include a $35,000 royalty payment as well as full documentation confirming the disposal of his unlawful soybean crop. Chancy, as well as the others involved, will make available all of their soybean production records, including Farm Service Agency/ASCS records, for Monsanto inspection over the next five years. Those involved will also provide full access to all of their property, both owned and leased, for inspections, collection and testing of soybean plants and seeds for the next five years.

Unfortunately, the Chancy case is not an isolated incident. To date, Monsanto has over 425 cases throughout the United States. Those cases were generated from over 1,800 leads received from ag-chem. retailers, seed dealers, growers and others. Over 200 of these cases currently are under investigation.... A sampling of some recently settled cases include the following:

• One Iowa grower will pay $16,000 royalty for illegally saving and replanting Roundup Ready® soybeans.
• Two Indiana growers will pay substantial royalties for illegally pirating patented biotech seed. One will pay $15,000, and the other will pay $10,000.
• In Illinois, a farmer will remit a $15,000 royalty payment for his unlawful actions.

Respecting these patents is more than a matter of avoiding legal risks. It is also a matter of protecting the development of future technologies. Monsanto has invested many years and millions of dollars in biotechnology research to bring our customers new technologies sooner rather than later. When growers save and replant patented seed, there is less incentive for companies to invest in future technologies that will ultimately benefit farmers. These technologies include seed that produce higher yielding crops, drought-tolerant crops, crops that are protected against rootworm damage, and high-value soybeans that may ultimately be used to produce plastic.

Following this 1998 letter, Monsanto became involved in another well-publicized seed piracy case, this time in Canada. On March 29, 2001, Federal Court of Canada Judge Andrew MacKay ruled in favour of the plaintiffs, Monsanto, against a seventy-year-old farmer, Percy Schmeiser, of Bruno, Saskatchewan. Monsanto had sued him for allegedly planting Roundup Ready® canola without a licence and was seeking the return of "all seeds or crop" containing the patented genes, and punitive damages for illegally obtaining the seeds, plus the company's court costs. Judge MacKay said that some of the remedies that Monsanto had sought should be granted, including monetary damages. He also banned Schmeiser from planting any more of the herbicide-resistant canola and from selling any of the crop or seeds from his 1998 crop.[40] For his part, Mr. Schmeiser countersued Monsanto for contaminating his fields with "volunteers," that is, plants that spring up from the seeds of last year's crops. In this case, the volunteers were not weeds but Monsanto's genetically modified canola.

Mr. Schmeiser's defence was as follows. For forty years he had grown oilseed rape every year, and he was therefore aware that something was unusual when he found canola, the new name for oilseed rape, growing around the electricity poles at the edge of his farm after he had sprayed the area with herbicide. Schmeiser realized that, among the seed he had collected from his previous crop, there were some GM seeds. By planting them, he had unintentionally contaminated his own crop, because if the GM plants were on the edge of his field, they were likely in the middle of it as well.[41] He claimed that he never knowingly grew Roundup Ready® canola. He argued that cross-pollination by the wind and bees, seed blowing off passing grain trucks or windblown swaths from another farmer's field could have made part of his 1997 crop contain the canola resistant to Roundup™.

Monsanto maintained a hotline so farmers could turn in neighbours for keeping and replanting gene-spliced seed, rather than buying it each year from the company, as their contract stipulated.

An anonymous caller accused Schmeiser, and Monsanto sent out Robinson Investigations to take samples of his canola crops. Part of Mr. Schmeiser's defence was that, since he had not signed a Technology Use Agreement with Monsanto, the investigators had no right to be on his land without his permission or to remove crop samples from it for their investigation. They also visited the local mill that chemically cleaned Schmeiser's seeds for planting. In both samples, the Monsanto gene was found.[42] Monsanto's lawyers pointed out that three separate tests performed on Schmeiser's canola showed that 90 per cent of it contained the patented DNA. Schmeiser's lawyer challenged the test's validity because it was conducted by Monsanto employees and said that a test performed by a University of Manitoba scientist showed a lower (but still significant) percentage of DNA, about 65 to 70 per cent.[43]

Schmeiser's day and a half on the witness stand in the Saskatoon courtroom enabled him to tell his story in his own words, but it also permitted Monsanto's lawyers to confront him with what they considered to be major anomalies in his account. After the defence lawyer finished his case, the opposing attorneys, acting on behalf of Monsanto, undermined the credibility of Schmeiser's only on-farm witness and raised the possibility that this hired hand had repeatedly told a local farmer that Schmeiser had planted Roundup Ready® canola.[44] Even after the court ruling against him, Schmeiser enjoys widespread support among those opposed to GMO technology.

Each of the more than a thousand farmers in Canada investigated for seed piracy by Monsanto since 1997[45] would add further detail to this legal, economic and social conundrum. For Monsanto, the issue is making a legitimate profit, and the legal and moral protection of their patented seeds and other products. For farmers in North America, it is the perception that their legal liability for reusing the seeds is unfair, and the resulting emotional distress when they are caught is intolerable. For farmers in India and other countries in the developing world, the bottom line can be life or death. In addition to seed piracy, other social issues regarding patents on hybrid and genetically modified agricultural products are also significant, especially as they relate to the developing world.

As reported in the *Guardian Weekly,* anecdotal evidence suggests that the farmers in Warangal district of Andhra Pradesh have serious issues with the new agricultural products marketed by agro-chemical companies such as Monsanto. On January 18, 1998, Nagarikani Yellaiah walked to his one-acre cotton field carrying a plastic bottle of insecticide, drank it, then lay down and died. By April of the same year, 350 farmers from the Warangal area had hanged themselves or drunk the poisons that had failed to save their crops.

Warangal is cotton country. Its dry land has been cultivated for centuries by subsistence farmers like Nagarikani
Yellaiah. The Warangal farmers had been growing conventional
cotton, but like millions of other farmers in India, they were
persuaded to plant hybrid crops. Ten years ago, few people in
Andhra Pradesh grew cotton exclusively. Most farmers collected
their millet, pulse and oil seeds, grew enough for their families
and sold what was left over on the local market.[46] The fact that
hybrid and genetically modified seed generally does not grow
as well when reused than when originally purchased shocked
these farmers.

The Green Revolution of the 1970s changed India from
being the world's largest importer of food grains into a self-
sufficient country. Thanks to the work of Nobel laureate Norman Borlaug and his colleagues at the International Maize and
Wheat Improvement Center in Mexico, new and high-yielding
strains of dwarf wheat were developed which transformed Indian agriculture. This was a project supported by the Rockefeller
Foundation. At the time, this support seemed to some in India
as unwanted American interference. But events proved that this
novel approach was justified by its results. Especially in the
Punjab, but in many other places on the subcontinent, a few
large regions took up farming on an industrial scale.

But this revolution reached Warangal late, and it arrived
with Monsanto and other companies, which advertised high-
yielding cottonseeds and powerful pesticides. Because modern
cotton farming requires expensive fertilizers and hybrid seeds
each year, the revolution brought the moneylenders as well.
Corporate promises, clever advertising and incentives persuaded
several million farmers to turn from their traditional crops to
the "white gold": cotton.

For some time, the Warangal farmers did well. However,
they were surprised when the world price for cotton fell, and
they were unprepared for the discovery that pests could build
up resistance to their chemicals. Also, the hybrid seeds required

more water than traditional kinds of cotton. "Death was the final solution," said Rameka, whose husband drank insecticide after his seeds failed. The first year was good, but later he lost money. "He borrowed the money to invest in a well,"[47] she added. Cotton is get-rich-quick farming, a bit like gambling. Once people are in debt, they must keep on growing it. If they can't pay their debts, in a moment of weakness, they may commit suicide. Rameka's husband owed $80 to a moneylender. Such a situation could perhaps have been avoided if Monsanto had been able to introduce genetically modified cottonseeds that had a built-in insecticide.

Monsanto has been making herbicides in India for 30 years. But since its move into genetic engineering, it has become a powerful player in Indian agriculture as a leader in the genetic revolution. Its bridgehead into the vast Indian seed market is cotton. Its product is "Bt (*Bacillus thuringiensis*) cotton" which is genetically engineered to resist, among other pests, bollworms, a worldwide problem facing cotton farmers. Its patented Bollgard™ seeds have been growing in China and the United States for several years. The company claims that they reduce the use of pesticides by up to 60 per cent and increase yields by up to 8 per cent, with a net gain of 30 per cent.

Monsanto wanted to introduce Bollgard™ into India in 1999, and had advertised its product widely. There had even been links made between Bollgard™ and religious celebrations such as Diwali, the Hindu New Year. However, Monsanto suffered a major setback as a result of its offer to purchase Delta and Pineland, a giant seed company that dominated the American cotton market. Delta and Pineland held U.S. patent No. 5,723,765. This patent, developed with the help of the U.S. Department of Agriculture, would cause a healthy plant to produce infertile seeds. The company felt that if the patented possibilities were realized, this development would preclude the legal and public relations problems associated with seed piracy. However, it was dubbed by the Rural Advancement Foundation International (RAFI),[48] a Winnipeg-based agriculture advocacy group, "Terminator" or "suicide technology." Monsanto and other multinational seed producers prefer the name "Technology Protection Systems" and are developing them to defend their plant breeders' rights, protected by several international agreements, which is to say, their patented intellectual property.[49]

In India, the very idea was obnoxious. In addition to the usual economic, political and intellectual objections to "Terminator," people in India added specifically religious ones. Religiously, seed has a mythic power for most of India's 500 million farmers because it represents cyclical renewal and the essence and means of life itself. Saving seed, reusing it and sharing it are considered fundamental freedoms that can never be surrendered. For these and other reasons, the Government of India introduced federal legislation originally intended, among other things, to ban the sale of terminator seeds.[50] That farmers should not be able to replant seeds was not thought to be a patent violation, but to be patently unjust.

As in many ethical dilemmas, both sides have strong arguments to make. Some insights from the religious past might therefore be of assistance in making a judgment on whether corporate intellectual property is an inviolable right, as Monsanto would claim, or whether it is a contingent right that must surrender to other social values.

3. Traditional Roman Catholic Thinking on Property Rights

The notion of private property has great importance for the Catholic Church's ongoing reflection on the concept of a just social order. The correct application of the concept of patents, the current growth of biotechnology in the global economy, and the many and growing inequalities between rich and poor people and nations, along with economic and technological advances, have made the idea of property, its moral foundations, its purpose and its limitations an important topic for discussion. The issue of private property has a relationship with just distribution of the world's goods, including food and medicine, as well as with essential human freedoms, such as those claimed by Indian farmers to save and reuse fertile seeds from the flower heads of crops grown with purchased seeds.

The concept of property differs legally and morally from one culture to another around the world. For our purposes, property is identified with ownership by a group of people who have purchased a large company such as Monsanto. It will stand for the thing owned such as patented and genetically modified Roundup Ready® canola and especially for the rights to its exclusive use exercised through Technology Use Agreements with the purchaser. Adapting a definition coined by a scholar who spent his whole life analyzing the evolution of the notion of property in modern society, it is, in essence, a relationship between a group of individuals (such as the corporate board and shareholders of Monsanto) and "a tangible or intangible thing."[51]

The magisterium, or teaching function, of the Catholic Church has unambiguously maintained throughout the centuries that there exists a natural right to possessions, whether these are tangible and movable or intellectual and immovable, whether they are productive or purchased property, whether one has permanent control of the substance or only partial ownership guarded by something like the Technology Use contracts that Monsanto expects its seed purchasers to sign.

This right is not inviolable. Pope John Paul II speaks of a "social mortgage" on all property and cautions that we face "a serious problem of unequal distribution of the means of subsistence originally meant for everybody, and thus unequal distribution for the benefits deriving from them."[52] He consistently employs as a foundation for his teaching on property the "universal destination of goods," which he calls "the characteristic principle of Christian social doctrine."[53]

This teaching is not new. Pope John XXIII put the matter clearly in his 1961 encyclical *Mater et Magistra*:[54]

Private property, including that of productive goods, is a natural right possessed by all, which the State may by no means suppress. However, as there is from nature a social aspect to private property, he who uses his right in this regard must take into account not merely his own welfare but that of others as well.

This paragraph encapsulates the classic argument of St. Thomas Aquinas (1224–74). In the Catholic conception of moral reality, all human rights flow from the nature of the human person. Included in these rights are rights "to bodily integrity, and to the means for the proper development of life."[55] In the Thomistic vision of justice, material things have come into existence in order to serve human needs. In fact, without their use, human beings cannot survive, cannot flourish through development of potential and talent, cannot fulfill essential obligations such as the care of one's family; in short, a human being cannot even practise virtue without the help of material things.[56]

It is in reality itself that the fundamental right to use material things is grounded. This is based on the intrinsic relationship between the nature of man/woman and the nature of material things. These grounds have to do with use of material things and not with ownership, but they seem to imply and concede the right to ownership as well, and this legitimizes the regime of private property.

A key question is whether this legitimization of private property is merely permissive, or whether it is intrinsic or of obligation. Aquinas argued that there are two different capacities of humans toward the material world. The first is to procure and dispense material things, and in this regard it is appropriate that we endorse the right to private property:

First, because every man is more careful to procure what is for himself alone rather than that which is common to many or all: since each one would shirk the labour and leave to another that which concerns the community, as happens where there are a great number of servants. Secondly, because human affairs are conducted in more orderly fashion if each man is charged with taking care of some particular things himself, whereas there would be confusion if everyone had to look after any one thing indeterminately. Thirdly, because a more peaceful state is assured to man if each one is contented with his own. Hence, it is to be observed that quarrels arise more frequently where there is no division of things possessed.[57]

The second thing that humans are capable of doing with regard to private property is to use it, and, he says, this use should also be for the common good and not only for private benefit.[58] The overall thrust of Aquinas' thinking should not be overlooked. His point of departure is the argument that appropriate accessibility of material goods is a necessary means for human self-fulfillment. The rights of an individual or ownership group are subsidiary to this basic principle. The right of use is what is absolute, primordial, and normative or regulative. Each man or woman, thanks to their personhood, is entitled to a share of the means necessary to his or her perfection. The right to actual ownership by the individual or a group of persons, on the other hand, is derivative and relative. It is to be discovered with the help of historical and sociological circumstances. Particular forms of property distribution differ, and the justice of each arrangement falls under the rubric of the universal right of the use of material things. In summary, according to the thinking of Aquinas, "the rights of property are to be respected but in the exercise of such rights the promotion of the common good and the function of material things to serve human needs and the destiny and dignity of man are controlling."[59]

4. Property in Scripture

Catholic thinking on private property belongs to reflection on the natural law. Because there is no clear answer to our question in sacred Scripture, theologians must employ what Thomas Aquinas called "right reason," our religiously informed intelligence, to solve the problem. We prefer to call this ethical methodology "faith-informed autonomous ethics." Despite the fact that Scripture does not generally give unambiguous answers to most moral or social questions, many indications found within its pages can give us guidance.

Perhaps the most obvious scriptural text legitimizing private property is the so-called Seventh Commandment in the Decalogue: "You shall not steal" (Exodus 20:15). But beyond this normative level of discourse, which would seem to be absolute, there is in the Hebrew Bible a distinctive conception of property in a community guided by divine revelation. In a variety of ways, the ancient Hebrews were reminded that what they possessed was held in stewardship from God and was to be used to promote an organization of equals appropriate to their calling as the People of God. This is probably the inspiration behind the prescriptions of the Sabbatical Year (Exodus 23:10, 21:2; Deuteronomy 15:12) that freed servants and debtors and subsidized the poor, as well as those of the Year of Jubilee (Leviticus 25:10, 23), that returned the agricultural property to its original owners every fiftieth year.[60] Israel had been trained in this notion of stewardship and sharing of food earlier during the Exodus from Egypt through the desert when the Lord fed them with manna. "This is the Lord's command: Everyone must gather enough of it for his needs ... according to the number of persons in your families ... They gathered it, some more, some less ... the man who had gathered more had not too much, the man who had gathered less had not too little" (Exodus 16:16-17). Moses instructed them not to gather any manna for the next day. "But some would not listen to

Moses and kept part of it for the following day, and it bred maggots and smelled foul; and Moses was angry with them. Morning by morning they gathered it, each according to his needs. And when the sun grew hot, it dissolved" (Exodus 16:19-21).

The prophets were outspoken in their condemnation of social injustice when greed of ownership reared its ugly head, dispossessing the weak (e.g., Isaiah 3:24-25, 5:8; Micah 2).

The New Testament's teaching on property is similar. Daily needs are a proper subject of prayer (Luke 9:3), but wealth has its dangers (Matthew 19:24). Jesus highly valued liberality and service to others (Matthew 25:14). The Gospel of Luke deals especially with questions of property; for example, Luke relates that Jesus refused to adjudicate an inheritance dispute (Luke 12:14) and used the opportunity to warn his listeners against covetousness instead. It is Luke also who relates the story of Zacchaeus, in which Jesus invited himself into the tax collector's home for the night. His repentant host gave away half his property to the poor as well as repaying any money embezzled from others at a rate four times the amount stolen (Luke 19:1-10). Jesus also described a man who had identified the good life with excessive wealth as a "fool" (Luke 12:20). He recommended voluntary abandonment of possessions in favour of the poor to a young man so that he could seek spiritual perfection (Matthew 19:21; Mark 10:17).

Do we have any guidance in Scripture about patents? A patent is a way to share the details of one's discovery with the rest of the community and, at the same time, be assured that this candour will not suffer the economic penalty of the non-rival use of one's invention or idea.[61] One interpretation of the parable of the talents (Matthew 25:14-30) leads us in this direction.[62] The servant, who hid his one talent in the ground, fearing that he might lose it, is severely condemned, whereas the master commends the other two servants who traded their talents to make other talents.

5. St. Bonaventure on the Goodness of Creation

A third source for Catholic theological reflection on our question comes out of prayer. This resource goes beyond the evidence gleaned by reflection on the natural law and Scripture. It does not contradict it, but simply deepens it. It is often the case that a theologian will share the insights of the founder of his religious community, as is certainly the case with St. Bonaventure (1217–74), a Franciscan. The religious experience of the founder of his Order, St. Francis of Assisi, had to do with an overwhelming conviction about God's overflowing goodness, and the experience of creation as an expression of that goodness. The whole notion of patents and patent legislation is usually discussed in terms of procedural justice. This is appropriate, but it is possible to offer additional perspectives on it as well. For this reason, it might be helpful to explore briefly how Bonaventure developed the mystical apprehensions of his founder.

Creation is, for Bonaventure, divine self-expression. In his Trinitarian theology, the Word, whom some theologians prefer to name the Son of God, is the internal self-expression of God's overflowing goodness. Creation, for its part, "is the external expression of the Word."[63] He often uses the name "Wisdom" to refer to the divine Exemplar whom creatures reflect. Creatures are, he writes, "nothing less than a kind of representation of the Wisdom of God, and a kind of sculpture."[64] In a remarkable passage, he waxes eloquent on how light coming through a window is reflected in the variety of creatures:

> As a ray of light entering through a window is colored in different ways according to the different colors of the various parts, so the divine ray shines forth in each and every creature in different ways and in different properties.[65]

6. Catholic Thinking on Patents

The theological poetry of Bonaventure reminds us that the good things given to humanity by God in nature belonged originally to everyone in common. At this original moment of creation, these fruits of the earth were not under the control or power of any particular person. Still, because God gave them for our use, there must be ways and means for us to claim them, to take ownership of them, so that they can be of use and benefit for us. This is a paraphrase of part of the seventeenth-century British philosopher John Locke's (1632–1704) argument that organizing the system of private property is a useful social contract.[66]

In much less mystical language than Bonaventure's meditation on sunlight passing through a stained glass window, economists call the visible light from the sun a "pure public good." This is because our enjoyment of it doesn't challenge any other person's, nor does our delightful use exhaust its bounty. It is "non-rival" and "non-excludible."[67]

Before the invention and development of the telegraph in 1837, the same could be said of all aspects of the electromagnetic spectrum, of which sunlight was the most prominent element. Until then, the rest of the spectrum had little or no practical use to humankind. Since then, technology has transformed other elements of the electromagnetic spectrum from unrecognized gifts from God and public goods to private assets of great value. Sunlight alone remains "non-rival," pouring through Bonaventure's church window and enriching his meditation, and ours, on the goodness of creation.

Radio waves were not known in the eighteenth century. Certainly they were not private property. They have since moved from being unrecognized blessings of God to public goods and finally to private property, through the application of science and the creation of laws and other social institutions. Through these laws, radio waves and the income streams they generate have become property. The nature of electromagnetic waves, especially their ability to move through infinite distances of space and the resulting applications in long-distance communication, means that effective laws governing their use must also cover vast distances. Telegraphy, telephone, radio, television and the new Iridium system that can transmit data and instant voice communications across the globe all need agreements on use and ownership.

As the telegraph spread from country to country, it soon became clear that its use would be inhibited without some standards and regulations. In 1865, a treaty signed by representatives of twenty countries founded the International Telegraph Union. The invention of the telephone in 1876 was followed by agreements on its use in 1885; the Berlin Radiotelegraph Conference followed the first voice radio transmission in 1920; the first broadcast radio in 1920 was followed by international agreements on its use in 1927.

With each technological innovation, new regulations had to be formulated. At the outset, ownership of these allocations seemed to have had little commercial value but, with the advent of advertisements, broadcasting licences soon became significant assets. Prior to 1837, the ether was universal and underappreciated; through the interrelated innovations in technology and institutions, it became private property.[68]

Conflicts have arisen. For instance, in 1963, the Hat Creek radio telescope belonging to the University of California unexpectedly detected a strong transmission that turned out to be the "first known natural maser, an intense blast of laser like, organized radio waves unleashed by molecules excited by cosmic radiation."[69] This very strong signal was located at 1612 megahertz (MHz), and has become a significant scientific tool that enables astronomers to estimate the diameter of stars and their distance from earth. While the maser is quite distinct and localized, any other signals close to 1612 MHz can interfere with the scientific applications. The astronomers sought protection of this point on the spectrum, and in 1992 the International Telecommunications Union awarded them primary rights to the spectrum from 1610 to 1613.5 MHz, to protect that area from any interference. However, the Iridium system has rights to an adjacent band, from 1616 to 1626.5 MHz, and stray radiation from Iridium has degraded the quality of the astronomers' signals.[70] Technological and regulatory solutions are being explored to resolve this property rights dispute.

The technology that makes use of recently decoded DNA to alter genes in order to transform and, hopefully, improve our food grains is very new. Already, a process to regulate its development has begun. The process parallels the regulation of the electromagnetic spectrum so that the dual aspects of the right to private property identified by St. Thomas Aquinas, personal benefit controlled by the common or social good, will eventually emerge.[71] The reason that all technology requires regulation and the patent system has developed is the value of the tangible and intangible, rival and non-rival goods involved.

Long before the patent system developed, the fourth-century Catholic theologian St. Augustine asked two questions about private property. The first was how one obtained property morally. The second was how one could continue to hold on to the property one had already obtained. The answer to the first question was that the legal right to private property is a fundamental civil right. The law of nature, as opposed to humanly created laws, is fundamental to the common ownership of property, but it is also the foundation of our right to private property. This is because Augustine thought that we each have a natural right to a fair share of what God has created for all.[72] When one turns one's attention to the second question, as to why we have a right to continue to possess property, natural considerations give way to the guidance provided by the divine law or human law.[73] The ethical acquisition and retention of property was, for Augustine, conditional on its just use. It was obvious to Augustine that honourable persons acting in conformity with the divine law must be using their property well in social terms.[74] They should, therefore, have the right to continue to hold on to their property.

With respect to unjust persons, the Emperor had the right to deprive them of their property at any time. Augustine's position was that the authority of the prince was absolute in the political, economic and social spheres of life. Augustine would allow that the Emperor has the right to confiscate the property of the upright in a time of emergency such as a famine or war. Failing these situations, a sovereign who was following God's law would have no excuse for demands on the right users of property other than taxation in order to maintain the system of government.[75]

In other words, Augustine saw that the use of goods was a "continuing creation" in co-operation with God's original creative act. Therefore, for him, it is "scarcity," rather than value, that is the basis of economics. Thomas Aquinas accepted this point of view and went on to say that the market value, or price, of a commodity depended upon its "aptitude to serve human uses" (its utility) as well as its relative scarcity. Antoninus of Florence (1389–1459) in his *Summa Moralis* (1477) argued that value had three components: (1) scarcity (*raritas*); (2) usefulness (*virtuositas*); and (3) desirability (*complacibilitas*).[76] Intellectual property related to our topic would be a new concept for these early Catholic theologians. But the basic practice of applying for and obtaining patents on one's inventions and discoveries would be congruent with their thinking.

That being said, today there is a growing consensus that our current thinking about patents and patent protection under the law needs to be revised in various ways.[77] At least sixteen patents are pending at the Canadian Patent Office that lay claim to not only physical DNA molecules, i.e., genes, but also to the digital representations of those molecules stored in computers. Any digital representation of a patented gene's chemical makeup, encoded as long strings of the letters A, G, C and

T stored in a computer, could be as much of a patent infringement as cloning, using or selling the actual DNA molecule.[78] An indication of the failure of the current patent system is that, because of widespread patent violations, Monsanto has opted to prevent farmers from reusing the seeds they have purchased by resorting to contract law rather than patent law. Farmers who signed the Technology Use Agreement, especially in developing countries, become indentured to the company. They surrender their freedom as farmers to plant their fields in their traditional way. Recognizing that the TUA protocols were not the ultimate answer to seed piracy, Monsanto and other companies then returned to a technological solution to the perceived injustice of seed piracy by developing Technology Protection Systems. A setback to this project occurred when the worldwide outcry against "Terminator" led Monsanto to make a public statement that it was abandoning this technology. In an open letter dated October 4, 1999, to Rockefeller Foundation president Gordon Conway, Robert Shapiro, who was then CEO of Monsanto, said:

> I am writing to let you know that we are making a public commitment not to commercialize sterile seed technologies, such as the one dubbed "Terminator." We are doing this based on input from you and a wide range of other experts and stakeholders, including our very important grower constituency.[79]

However, Mr. Shapiro went on to say in the same letter that they would exercise their right to develop other Technology Protection Systems, such as Genetic Use Restriction Technologies. Also known as GURTs, this genetic modification, which is still at the research stage, would allow a seed consumer to "switch on" desired genetic traits in a seed by using a proprietary spray that would have to be purchased from the seed company. If the spray were not applied, the seeds would grow, but the genetically modified trait or traits would not appear. The rationale for GURTs is the same as for "Terminator": patent protection and making a fair profit on the heavy initial investment by the company and its stockholders. The worldwide protest against "Terminator" won a victory, but it was a pyrrhic one.

7. The Socialization of Patents

Pope John Paul II has repeatedly spoken about the pressing need to understand the notion of universal love in terms of solidarity. In itself a virtue, solidarity is "a firm and persevering determination to commit oneself to the common good; that is to say to the good of all and of each individual, because we are all really responsible for all."[80] In a Jubilee Year address directed to those involved with agriculture, he said: "...we need a *globalization of solidarity*, which in turn presupposes a 'culture of solidarity' that must flourish in every heart."[81] On the one hand, patents would apparently represent a contemporary example of an area where the "virtue of solidarity" does not operate, but on the other hand there are indications that this is changing.

Patents represent a compromise between the individual right to property and the social obligations incumbent upon all property owners. A patent is a time-bound, negative right: that is, by issuing a patent, government officials have made it illegal for anyone else to use or profit from this invention without the consent of the patent holder. It does not constitute a positive right to make use of the patent nor does it amount to an ethical agreement concerning what is patented. Until it expires, those who wish to use the invention are obliged to pay the legal owner or otherwise obtain permission. The details of the invention are available for all to inspect in public documentation. After twenty years, or whatever the legal life of the patent is in a particular country, the discovery is placed in the public domain and belongs to all.

The legal protection of and just remuneration for intellectual property motivates innovation and creativity. Countries with enforceable and enforced patent and trademark legislation often enjoy economic benefits. "The evidence suggests that intellectual property protection is a significant determinant of economic growth."[82] Economists today stress that "[e]xplanations of economic growth are increasingly focusing on the power of expected profits to motivate innovation."[83]

The bioscience and pharmaceutical revolution that has followed the decoding of the genome has tended to diminish the social aspect of patents and emphasize the private ownership dimension. For instance, because of the technical possibilities of DNA technology, companies have applied for and been granted patent rights to biotechnological research methods. "No one should be able to patent a gene or any genetic chemical sequence unless they know what it does and what its potential uses are."[84] In the U.S., the requirements for patents are the following: that the discovery is something that the law allows to be patented; that the invention will be "useful"; that it is possible; that a written description is supplied; that it is "novel" or "non-obvious." This permits the patenting not only of transgenically produced plant traits and genes, but also of crucial genetic fragments needed to perform all genetic modifications.[85] Because so many of the procedures and methods that are useful for research into transgenetically modified food are patented, countries that very much need to grow more food and to develop their pharmaceutical industries through "biopharming" may not be able to do so.

An analogy might be suggested between the way patent law treats pharmaceutical products delivered through transgenically modified agricultural products and the way it treats methods of medical treatment. Even though life-saving surgical techniques might profit their developers, the law does not allow them to be patented. The European Patent Convention, the Magna Carta of European patent protection, states in Article 52(4) that "methods for treatment of the human or animal body by surgery or therapy and diagnostic methods, practised on the human or animal body, shall not be regarded as inventions which are susceptible of industrial application."[86] The reason for this exclusion would seem to be the need to respect the social effect of patents.

One response to this omission of surgical techniques and other life-saving medical methodologies would be to amend future European patent legislation to include them. In fact, their absence appears to be an anomaly, in view of the fact that life-saving pharmaceutical discoveries are subject to patents. The correct response would be to see this as a directive towards a more humane approach to the future protection of intellectual property.

Is the fact that potentially life-saving medical techniques are not (yet) understood as patentable to be seen as prophetic? Catholic theology has extended the biblical notion of an ancient prophet commenting on the religious implications of a contemporary situation to include present-day events and their implications, especially as these relate to social justice. By doing this, theologians intend to make the claim that the insight and consequent understanding about the human condition and human responsibility that ground this social decision come indirectly from God as a pointer towards other, analogous choices that must be made in the future. If life-saving techniques such as the Heimlich manoeuvre, which has saved countless people from choking to death, cannot be patented, should not the same resolution be extended to life-saving drugs?

A second prophetic pointer to Monsanto and other seed companies facing decisions on how to protect their intellectual property rights and to do justice to their stockholders came from another American giant in the biotechnology field, Merck. In 1987, Merck Chairman and CEO P. Roy Vagelos, M.D., announced that he and the other stakeholders in his company, after a long and difficult debate, had decided to donate their patented drug Mectizan™ to all who need it for as long as necessary to treat a disease called river blindness that is caused by microscopic parasites. Now in its fourteenth year, the Merck

Mectizan™ program treats 20 million people in West Africa who would otherwise become blind. Quite apart from the impressive example of corporate philanthropy that this represents, indications are that it has been helpful to Merck's public image and has not adversely affected the exercise of its patent rights to this drug. People and organizations that are interested in having a "social investment" in their stock portfolio would insist on including Merck precisely because of its use of the intellectual property of Mectizan™.

The Board and stockholders provided evidence of their continued support through their 1998 decision to allow Merck to begin treating a much larger population of 300 million in Africa at risk for a similar problem, lymphatic filariasis (elephantiasis), with the same patented drug in combination with either albendazole or diethylcarbamazine (DEC). The results of this new treatment are remarkable, reducing the parasites' level in the blood by 99 per cent.

Onchocerciasis, or river blindness, is caused by the bite of adult worms (macrofilariae), which release minute worms (microfilariae) into the victim's skin. These minute worms cause itching, depigmentation of the skin and eventually blindness. Onchocerciasis is endemic in West Africa but is not found elsewhere. Prior to its control, 34 million people were at risk of becoming blind from this disease and millions of hectares of fertile land were deserted out of fear of contracting it through exposure of bare feet to the waterborne parasite in irrigated fields.

Microfilariae can be treated by ivermectin, but the treatment must be continued for about fifteen years. Prior to its decision to donate Mectizan™ to the Onchocerciasis Control Program of West Africa, the company had to debate the following question: what price, if any, should the company charge for Mectizan™?[87] Residents of some of the poorest countries in the world are the only people who suffer from river blindness. They are among the least able to pay the market rate for the drug, a price that would compensate Merck for the cost of its development and provide a profit to its shareholders. However, the company also had to ask whether a decision to donate the drug would create an expectation of further philanthropy in the minds of victims of other diseases. Would this choice then become a disincentive for other research against tropical diseases? And, in addition to the cost of research, development, manufacturing and administering the Onchocerciasis Control Program in eleven West African countries, would the company face legal liability if unexpected adverse reactions should develop among those who used their drug?

Ultimately, the company's decision was based on a simple credo articulated by George W. Merck, the company's president from 1925 to 1950. In a 1950 address to the Medical College of Virginia, he said, "Medicine is for people. It is not for profits. The profits follow, and if we have remembered that, they have never failed to appear." Later in the same speech, Mr. Merck said, "How can we bring the best of medicine to each and every person? We cannot rest until the way has been found with our help to bring our finest achievements to everyone." Based on this company credo, Merck decided to give their medicine away for free throughout the world. Without Mectizan™, the research scientists and other employees at Merck knew that millions of people would continue to lose their sight.[88]

The prophetic stance taken by Merck, and its social benefits both to the company and to the consumer, were not lost on other drug companies. But prophecy in a global economy is never without its money-making aspect.

Early in 2001, a spokesperson for the New York-based drug company Bristol-Myers Squibb announced that it had come to an agreement with the World Health Organization and the medical charity, Doctors without Borders, to supply Eflornithine to Africa for free for the next three years. On the surface, this free gift seems to lack congruence with the normal practices in a market economy other than those of Merck, described above.

But the CEO and the members of the Board of Bristol-Myers Squibb have an easy answer for stockholders who might question the worldly wisdom of generosity to Africans suffering from sleeping sickness.

Eflornithine is effective against sleeping sickness (trypanosomiasis). This illness, endemic throughout Africa, is contracted through the bite of the tsetse fly. The saliva of this small fly carries a disease-causing protozoan parasite from an already infected person.

The symptoms of trypanosomiasis start with a mild fever, itchy skin, joint pain and lethargy. But weeks later, when the parasites infect the brain, patients begin hallucinating and acting in ways dangerous to themselves and others. Their skin becomes so sensitive that their caregivers cannot wash them with cool water or even touch them without causing them to scream out in pain. The name "sleeping sickness" comes from the fact that towards the end of their lives they fall into a lassitude so profound that they cannot even feed themselves, which leads to death.

Because the fly tends to live in dense brush near water, the illness is a particular scourge for women and the babies they carry on their backs as they get water or wash clothes.

Eflornithine, also known as DMFO, was one of only four drugs known to work against sleeping sickness. It was originally developed by Aventis and sold as an anti-cancer drug marketed under the name Omidyl. Its usefulness against sleeping sickness was discovered by accident in the 1980s — it was so effective, it became known as the "Resurrection Drug." Unfortunately, two factors militated against its use in Africa: it was expensive, and it proved to be ineffective against cancer. The latter circumstance meant that it could not be sold in the West. It was not needed in the West against sleeping sickness, which is not found there, so there was no commercial market. Aventis, which still holds the patent on Eflornithine, stopped production in 1995 and handed over the production licence to the

World Health Organization (WHO) in 1999. In that same year, the company found 227 kg of a precurser chemical and made one batch of 7,800 vials, which was to have been the last use of this remarkable drug.[89]

This was by no means the end of the story of Eflornithine. After the last vials of the life-saving drug were exhausted, Bristol-Myers Squibb discovered that the drug had another, money-making use – reducing women's facial hair. A deal struck with the WHO allows it to produce Vaniqa™ – a prescription-only cream that slows the growth of facial hair, in exchange for 60,000 doses per year to provide treatment across Africa, beginning in May 2001. In return, the WHO will allow Bristol-Myers Squibb to produce and market the cream.[90]

There are other examples of the social uses of patents and production licences, as well as instances of collaboration between pharmaceutical giants and the World Health Organization. For example, there is the announcement by GlaxoSmithKline of a joint project with the WHO to develop a new malaria treatment, Lapdap, which will be sold cheaply to poor countries in Africa.[91] These examples are germane to our subject of biotechnology and food grains because the transgenic process may enable drugs to be delivered to those who need them by genetic modification of food. This has already begun to happen.

In early 1999, Professor Ingo Potrykus of the Swiss Federal Institute of Technology in Zurich and Peter Beyer of the University of Freiburg in Germany created "golden rice," with financial assistance from the Rockefeller Foundation. In addition to the genes of *Oryza sativa*, one of the most common forms of rice, each grain contained pieces of DNA from bacteria and daffodils that would alter it to include vitamin A. Among potential consumers are "at least a million children who die every year because they are weakened by vitamin A deficiency and an additional 350,000 who go blind."[92]

The genes that Potrykus and Beyer transferred into *Oryza sativa* and the bacteria they used to transfer them were all encumbered by patents and proprietary rights. In April 2000, the two scientists struck a deal with AstraZeneca (now part of Syngenta Ltd.), which holds the patent on one of these genes. In exchange for commercial marketing rights to "golden rice" in the U.S., Canada and other affluent countries, AstraZeneca dedicated itself to the cause of putting the seeds into the hands of poor farmers at no charge.[93]

Not to be left behind, by August 2000, Monsanto had made a similar undertaking with regard to its patents on elements of the way that "golden rice" is transgenically produced. Potrykus said in response, "I consider the Monsanto offer important because I can now use this case to tell other companies, 'Look, Monsanto is giving me a free licence. Won't you do the same?' It's an important first example."[94]

Critics of transgenic modification of food are suspicious of these developments. They often interpret them as instances of self-serving public relations by a very powerful industry that is faced with a costly consumer backlash against its other products.[95] There is some truth to this. At the same time, one can say with the Letter of James: "Every generous act of giving, with every perfect gift, is from above, coming down from the Father of lights, with whom there is no variation or shadow due to change" (James 1:17). Obviously, the multinational biotechnology corporations differ from "the Father of lights." But if there is any goodness in their apparent generosity to the poor, this goodness comes from God. Within the morality of intellectual property rights there is no "perfect gift," and all of human life, but especially the world of biotechnology, is subject to injustice and deception. In this respect, therefore, biotechnology and its patent system are far removed from the "Father of lights." It is possible to hope that the exclusion of medical procedures from what Europeans understand to be subject to patent applications, the generous philanthropy of Merck with regard to Mectizan™, and the waiving of patent rights by AstraZeneca and Monsanto on "golden rice" are all signs that the things, ideas and procedures known as intellectual property ultimately belong to all, especially those with the greatest need.

In 1477, eighteen years after his death, the first and certainly the most comprehensive treatment from a practical point of view of Christian ethics, asceticism and sociology in the Middle Ages was published by the editors of St. Antoninus (1389–1459). Though largely forgotten by Catholic moral theologians today, his *Summa Moralis* has pride of place in our discipline between the writings of St. Thomas Aquinas and St. Alphonsus Ligouri (1696–1787).[96] In the fifteenth century, he taught that the state had a duty to intervene in business affairs for the common good, and to give help to the unfortunate and needy. He was among the first Christian moralists to teach that money invested in commerce and industry was true capital; therefore, it was lawful, and not usurious, to claim interest on it. He still considered usury a sin, a position he shared with all moralists at that time and earlier,[97] but taking interest on capital was not necessary usury. "The object of gain," wrote Antoninus, "is that by its means man [sic] may provide for himself and others according to their state. The object of providing for himself and others is that they may be able to live virtuously. The object of virtuous life is the attainment of everlasting glory."[98]

The distinction between usury and legitimate profit on a capital investment first articulated in the Church in the sixteenth century is crucial for the moral evaluation of biotechnology. Antoninus was known in his day as an outstanding counsellor, as well as a reformer of the Catholic Church. All his books were of a practical nature, and in them he faced the most difficult issues of his day. Today the same need exists in the Church. Its theologians must insist that the right to private ownership of intellectual property has two moral purposes: it is a means of protection for personal self-determination, and it is also a means by which society is organized to fulfill basic human needs.[99] Social justice insists that our system of patents and other legal provisions be directed and structured in such a way that both of these purposes be achieved.

What makes this position challenging is that patents and the academic protocols that require scientists to credit others for their contributions are challenged by our sinfulness. The temptations of financial gain, fame and academic glory lead to moral challenges. Social justice in this area is a precarious achievement, only possible with the help of divine grace.

Chapter 3

Risks of Genetically Modified Organisms (GMOs)

Petitio principii is the Latin term for the most common error in logic. Commonly translated as "begging the question," this fallacy assumes something to be true, moral or good that has not yet been proven to be so. The current discussion of intellectual property rights would fall into this problem if it assumed that the novel food produced by DNA technology carried no risks. This is not the case.

In other words, one should not presuppose that GMOs are moral and then adjudicate ownership issues. While patents give their owners negative rights over their intellectual property, and Catholic thinking would generally support these rights, the actual production of GMO food products is a separate question. There may be risks to consumers if the owners of the patents decide to claim a positive right to produce genetically modified food, as they have been doing for the past seven years. This is especially the case if they did not identify these risks, if any, on suitable labels.

Catholic moral theology in previous centuries developed a conceptual language called "probabilism" which may undergird the cautious willingness of the Pontifical Academy for Life at the Vatican to give the new technology the benefit of the doubt with respect to risks. Using a different language and thought system, the World Council of Churches would tend to describe this technology as a prime example of human pretension with respect to God's original creation, and to see it as fraught with risks.[100]

After highlighting some public opinion surveys to demonstrate that there is consumer confusion over the new technology, I will describe two types of risk. The first has to do with the very process used to produce the novel food. The second has to do with specific and discrete products that result from this process. After adumbrating various risks that have been identified and those that may still await us, I will engage in another excursus into history, this time into the seventeenth- and eighteenth-century discussions by Catholic moralists, not of biotechnological risks, but of religious ones. In conclusion, I will make the claim that the foreseen benefits of genetic engineering outweigh the known risks.

1. Public Opinion

Canada is the world's third-largest producer of transgenically enhanced crops, after the U.S. and Argentina.[101] According to a Health Canada study done in the summer of 2000 by Environics, a public opinion research organization, if biotechnology can be shown to be beneficial, especially in the treatment of diseases, Canadian consumers are willing to accept minor risks. "When it comes to making decisions about the management and control of biotechnology products," the study says, "a majority of Canadians see scientific evidence as more crucial than people's concerns and perceptions."[102] This indicates a grow-

ing, albeit conditional, tolerance in Canadian consumers towards risk management by regulators and the scientific community. At the same time, Canadians indicate a strong disap-
proval about the way the food industry and commercial farmers
have foisted the new biotechnology on them. In an earlier study
conducted for Industry Canada by the same company, 75 per
cent of all Canadians were shocked to learn that three-quarters
of their store-bought food was genetically "enhanced" and that
this had been done without their knowledge or consent.[103]

As mentioned in the introduction to the book, perhaps the
most significant market research thus far on our questions has
been done by Ipsos Reid, another Canadian-based organization
dedicated to the scientific study of public opinion. Ipsos Reid
did an international survey in conjunction with the British journal *The Economist* about six months before the Environics' work
was completed. Among the Canadians questioned, 68 per cent
of consumers said that they would be less likely to buy something if they knew that the product had been genetically modified, or contained genetically modified ingredients. The same
wariness result, 68 per cent, was elicited throughout the world,
but in Germany 82 per cent of the grocery shoppers said that
they would have to think twice if a label showed an item that
contained ingredients that had been subjected to gene splicing;
in the UK the result was 67 per cent.[104]

A similar pattern emerges if one studies and compares consumer attitudes toward GMOs in various cities in Canada. In
January 2001, Winnipeg-based Prairie Research Associates interviewed 1,500 adults from four Canadian cities: Winnipeg,
Halifax, Quebec City and Edmonton. They have found that 46
per cent of Winnipeggers think GM foods are not as safe as
other foods. This is a greater percentage than those surveyed in
1999 when 41 per cent expressed misgivings around potential
risks. Among those surveyed, respondents who said that the
GM foods are safe increased 2 per cent from the 1999 survey:
26 per cent of people in Winnipeg and 27 per cent of Canadi-

ans overall think that the novel food is safe. Among other results of this study are the following: 94 per cent of Canadians want all GM food to be labelled; 78 per cent believe that Canadian meat found in grocery stores is safe; and more than half of all Canadians, 54 per cent, are concerned about eating GM foods.[105]

These opinion poll results confirm scientifically what we already suspected. Genetic engineering is a revolution that has overtaken the consumers of the world without their full awareness or consent; they are angry and confused, and they have real concerns about the risks involved. In some parts of the world, Europe and India in particular, these negative feelings have led to mass public street protests and boycotts by consumers. These actions have in turn resulted in government measures to regulate the domestic biotechnology industry and to restrict GM imports from other countries. In the mind of the ordinary person, the farmer and the corporate owners of giant multinational companies providing them with their transgenically modified seeds and chemicals are the ones to profit from the new advances.

The development of creative ways to provide life-saving drugs cheaply through genetically modifying food, as well as to deal with environmental issues through bioengineering of trees and plants, may help to change this consumer perception. For instance, let us revisit briefly our earlier discussion of "golden rice," which has been engineered to produce beta carotene, a chemical converted to vitamin A by humans.[106] This could be "good news" in the Christian sense of "Gospel" for the 500,000 children who go blind each year due to vitamin A deficiency, not to mention the two million who die from the same cause. Scientists at Cornell University in the United States are developing a banana that contains a hepatitis vaccine. Health care workers in the developing world may soon be able to treat their patients with this vaccine at a cost of two cents per dose versus $125 (US) for an equivalent injection.[107] Similar work is also underway on potatoes that express edible hepatitis B vaccine.[108] Mercury poisoning in the soil might be reduced, thanks to the development of a transgenic version of the common poplar.[109]

Faced with these risks and benefits, informed consumers are subject to moral complexity as they try to resolve their own consciences regarding the issues presented by biotechnology. If we ask what resources Catholic tradition could offer them in helping them to resolve their ethical conundrum about whether to purchase and use GM food for themselves and their families, some guidance might be offered from an unusual direction: the seventeenth- and eighteenth-century discussions in Catholic moral theology about the correct application of the practice of probabilism by confessors trying to help their penitents to resolve their conflicts of conscience.

2. Labelling

Much of the following discussion of probabilism depends upon the practice of labelling GM products, or, at least, stating which ones were "GM free."[110] The 2001 Royal Society of Canada recommendations for the regulation of food biotechnology were critical of the Canadian government's failure so far to protect the public from the risks of genetically modified foods and other biotech products. However, the Society fell short of recommending mandatory labelling of GM food.[111] The first steps towards a global consensus on the need to forewarn consumers through labelling seem to have been taken, despite the many problems involved. At the January 2000 World Trade Organization meeting in Montreal, 140 countries, including Canada and the United States, reached a compromise agreement that would allow countries to ban the import of a genetically modified food without full scientific proof that it was unsafe. This would require a label on bioengineered export crops that says they may contain

GM products. This agreement, which was unexpected because of earlier opposition from the U.S., Argentina and Canada, could speed the labelling of genetically engineered food in the local grocery stores.[112] At the time of writing, the presence or absence of genetic modification in our food is a matter of informed guesswork.

To illustrate such consumer conjectures, let us go through the lineup of a fast-food restaurant in Canada or the United States. On the grill are chicken and beef burgers. Like people, chickens and cows are vaccinated regularly to prevent disease, using recombinant DNA. This greatly reduces the need for antibiotics, which is good for the consumer, but not if there is risk in the way that these vaccines have been produced. To add further complications to the question, it is likely that these animals were fed corn and soybeans produced using crop biotechnology. Then consider the buns and toppings on your burgers. Through biotechnology, we can make improvements within a species that would otherwise require years of plant breeding. For instance, we can take a gene from winter wheat and add it to another type of wheat in order to enhance its protein content. This may have happened to the bread used in the restaurant. The tomato that lies on top of the meat in the bun may contain a reverse copy of one of its own genes that helps to slow its ripening process. The burger comes with french fries. They were probably cooked in canola oil, which is good for the hungry customer because it contains unsaturated fatty acid and tastes good. However, if there is a risk in the transgenic process producing the canola, the same french-fry lover should know that currently about 65 per cent of the Canadian canola crop is produced through biotechnology. The percentage is about the same or even higher in the United States. In addition to fries, one's order may include a soft drink such as a cola. Most soft drinks use fructose as a sweetener. Fructose is derived from corn, about one-third of which is produced in Canada and the United States using biotechnology.[113] Of all the potential risks in the restaurant, the genetically modified tomato offers the least because it has undergone exhaustive tests. A closer look at the history and process of the transgenic modification of one type of tomato might be helpful at this point in the discussion.

3. Testing

The Flavr Savr™ tomato was one of the earliest examples of a genetically modified food. In Canada and the U.S., most tomatoes are harvested while still green and firm, which gives the producers and marketers time to ship, handle and deliver them to the fast-food restaurant before they rot. In the past, they were artificially ripened by a chemical spray that released ethylene, a natural plant hormone. Unfortunately for the would-be customer, this process did not enhance the flavour. Although the tomatoes looked red and ripe, they were hard and tasteless.

A gene called polygalacturonase (PG) causes natural ripening in the tomato when it degrades cell wall pectin. Calgene, a California-based biotechnology company, isolated the DNA which codes for PG. Then researchers genetically reversed a copy of it, thereby producing an "anti-sense" gene, which they reintroduced into the tomato through the method of gene splicing. The anti-sense PG DNA produced a messenger RNA inside the nucleus of the cell that had the effect of inactivating the natural ("sense") tomato PG gene. The effect was to reduce greatly the amount of PG enzyme produced by the tomato, slowing the rate at which the tomato would ripen. As a result, the tomatoes could stay on the vine longer and acquire more flavour. There was still time to ship them to the supermarkets as well as to chicken and hamburger restaurants.

In 1989, although there was no requirement for regulatory approval at the time, Calgene began voluntary studies under the direction of the U.S. Food and Drug Administration (FDA) to assess the safety of their tomato. The CEO and members of the Board of Directors of Calgene believed that consumer opinion and acceptance would only support their novel food "with full, transparent, informed, and scientifically objective scrutiny – to give customers confidence and the information on which to make their choice."[114] The case became a regulatory benchmark, not only because it was the first genetically engineered whole food to receive FDA scrutiny and then reach the market, but also for setting the guidelines for FDA applications for transgenic foods. "The FDA concluded that Flavr Savr™ tomatoes did not differ significantly from traditionally bred tomato varieties. They were seen as functionally unchanged except for the intended effects of the anti-sense gene...."[115] Therefore, these genetically modified tomatoes were judged to be food and were subject to regulation as food. At the time, even critics of biotechnology urged the FDA not to impose such prolonged, strict and cumbersome procedures as were directed at Flavr Savr™ and Calgene during its five-year voluntary review. Subsequent cases in the U.S. have moved faster and have involved a different type of investigation.

The first Flavr Savr™ tomatoes were sold in Chicago and California on May 30, 1994. In 1997, Monsanto, which had by then taken over Calgene, withdrew the Flavr Savr™ tomato from the U.S. market because the original variety chosen for the genetic modification was not a good one for taste, texture and disease resistance.

The FDA studies of the Flavr Savr™ seemed to provide the scientific community with assurance that the process that produced transgenically modified food was safe. This was the calm before the storm. In Britain and elsewhere, the public began to panic over safety issues, and critics of GM foods began calling them "Frankenfood," after Frankenstein's monster. One reason for the panic was the publicity surrounding the research of Dr. Arpad Pusztai at the Rowett Institute for Agriculture in Aberdeen, Scotland. At a press conference in 1998, Dr. Pusztai announced some negative preliminary results from feeding rats a diet of potatoes bioengineered with a certain class of protein. No one involved with the potential toxicity of tubers, whether genetically altered or not, was surprised at this. Dr. Pusztai's experiment had no immediate implications for human health because there were no plans to feed these potatoes to humans. But Dr. Pusztai and his colleagues also argued that the rats were harmed not by the protein but by the method of inserting it into the potatoes. Because this method is the same as the one used in the Flavr Savr™ and its successor food products, the public and the British media began to panic over food safety.

Dr. Pusztai's potatoes had been engineered to produce a molecule called *Galanthus nivalis agglutinin* (GNA), a natural insecticide found in snowdrops (*Galanthus nivalis*). Dr. Pusztai's research hypothesis was that the molecule might make potatoes resistant to aphids. Speculations about the safety of the process that produces transgenic modification of food was fuelled by reports of damage to the rats in the form of thickening of the gut lining and poor development of organs such as the kidneys and spleen. The control group of rats, which had eaten ordinary, unmodified potatoes, suffered to the same degree as the experimental group stunted growth and suppression of the immune system, but did not have the other problems.

Dr. Stanley Ewen, a pathologist at the University of Aberdeen, suggested that a strand of genetic material known as the 35S cauliflower-mosaic-virus promoter might be responsible for the difference in outcome. Promoters are DNA switches used to turn genes on and off. Every gene has at least one of these, but it is possible to replace a gene's natural promoter with another that is more amenable to bioengineering. The 35S promoter is popular in biotechnology, and is found in a number of genetically modified crops including Monsanto's Bt-corn and Roundup Ready® soya. The promoters can end up in the wrong place in a chromosome, and begin switching on the wrong genes. This, according to Dr. Ewen, could have accounted for Dr. Pusztai's observations.

Other scientists, such as Maarten Chrispeels from the University of California, were skeptical. Potatoes are full of toxic compounds, which vary widely in concentration, depending on how the potatoes are grown. This phenomenon, known as somaclonal variation, makes feeding potatoes to laboratory animals particularly complicated as a scientific experiment. In other words, giving the same amount of potato to the test animals and to the controls will not produce reliable results. The potatoes themselves must have been cultivated in a uniform way to make sure that the chemistry in the tubers does not vary. Dr. Pusztai's experiment did not do this.

Over the years, scientists conducting experiments involving the 35S promoter have not reported any observations similar to those of Dr. Pusztai when he used genetically modified potatoes.[116] However, research done by biotechnology industries is not always open to public scrutiny, due to confidentiality agreements with the scientists involved. It might or might not raise such concerns. Despite the publicity surrounding Dr. Pusztai's experiments, at the end of the day, all his scientific colleagues can conclude is that rats don't like his potatoes!

4. Two Types of Risk

The fact that many scientists reorganized their lives and work schedules in order to address Dr. Puzstai's flawed experiment points to the fundamental importance for food safety of the *process* by which genetic modification is done. Opponents of mandatory labelling point out that it is not clear whether the issue of risk has to do with the bioengineering process that has produced the food or the end product itself. If it is the former, an obligation to label all GM products might prove harmful to the food industry, to commercial farming, and to the future benefits of biotechnology. This is because all of its products would potentially be harmful. If it is the latter, safety benefits might result.

An example of the process issue in labelling would be the MSG controversy that occurred in Indonesia and Malaysia in 2001, although it does not involve transgenic modification.

Monosodium glutamate (MSG) is a synthetic seasoning. Once widely used in Western cuisine, it waned in popularity when consumers became concerned about a variety of health issues, including the possibility that it was carcinogenic. However, it is still widely used in Asia. A major Japanese producer of MSG and other food products, Ajinomoto, faced a consumer boycott and other problems among Muslims in 2001 because one of the enzymes used in the catalytic chemistry that produces MSG is derived from pigs. It was alleged that the pig enzyme violated Islamic beliefs that prohibit the eating of pork. The fact that Ajinomoto marketed this product in Indonesia and Malaysia, most of whose citizens are Muslim, precipitated a crisis during which Indonesian police arrested eight employees of the company.[117]

After removing hundreds of tons of its product from Indonesian grocery shelves, Ajinomoto officials responded to the allegation. They pointed out that, although a pig enzyme was indeed used in the production process, it was merely a precursor element to enable the chemical synthesis to begin. The end product, the MSG that one applies to the food, does not contain this enzyme.

In Malaysia, labelling became a key issue. Authorities seized 80 kg of MSG believed to have come from Indonesia after the scandal of the pig enzymes had broken. The MSG was then repackaged, with labels in Chinese and English that did not make any claim that the contents were *halal*, that is, permitted under Islamic law.[118] The regional director of the Malaysian Domestic Trade and Consumer Affairs ministry told an Associated Press reporter that Muslims in Malaysia could be misled because all Ajinomoto products marketed in Malaysia are required to carry the *halal* label. In any event, consumers assume that any processed or manufactured food that they purchase in

Malaysia is *halal*. They do not read the label to confirm this and, in the case of the Ajinimoto MSG, were outraged when they discovered the truth. However, to demand that Ajinimoto state that the catalytic process used to produce the product included pork enzymes would lead to further marketing trouble for the company.

Although the question of whether the process used by Ajinimoto MSG is acceptable remains unresolved, the eight employees of Ajinomoto were subsequently released and the Indonesian government has declared that their MSG contains no ingredients outlawed by Islam.

An example of the second potential purpose for labelling, to warn consumers that a particular product is potentially hazardous to their health, would be the unfolding tragedy of the sheep disease called scrapie, the cow disease Bovine Spongiform Encephalopathy (BSE), and the human one known as new version Creutzfeldt–Jakob Disease (vCJD) that threatens the world's use of beef as food. The BSE crisis is also instructive as a partial explanation of why Europeans, especially the British, are concerned about the risk of genetically modified food. The MSG controversy contrasts with the BSE/vCJD situation in Europe because of the difference between a moral risk, as in the case of MSG in Muslim Asia, and a health risk with BSE. The MSG case represents a challenge for those who would demand full disclosure on labels, and the risks involved in the possible contamination of a food product. Large agro-businesses would not want to see a popular outcry begin, nor would they want to have their employees arrested and detained by civil authorities. Despite these concerns, these businesses may have to face the prospect of mandatory labelling of their consumer food products. The BSE example that follows, in which humans were infected by a dangerous disease, suggests that public health and regulatory officials have a prima facie moral obligation of full disclosure.

In the words of one commentator, the publication in October 2000 of the results of the BSE Inquiry in Britain represents "a case study of how science works, how scientists and politicians interact, and how governments cope with uncertainty and changing public attitudes to risk...."[119] Subsequently, BSE broke out unexpectedly in early 2001 in thirteen other European countries.[120] This included an outbreak of the human form in France. As a result, the story is ongoing, and is full of strong visual images and poignancy. Those who saw it will not soon forget the live television pictures of a failed publicity effort by the former British Minister of Agriculture, John Gummer. He is shown trying unsuccessfully to make his daughter Cordelia eat a very hot hamburger. The viewer wonders whether her rejection of it is prescient, wise.

The *New York Times* brought the plight of the cattle industry home to readers with its description of the French cattle breeder Guy Jambon, whose entire herd was slaughtered in February 2001 after a single case of mad cow disease was discovered on his farm. After all 670 of Mr. Jambon's cows were taken for slaughter and incineration, his barns, pens, the milking room and the fields were all empty. "The part that got to you was how quiet it was," Mr. Jambon said. "I was saying to myself, 'We don't really understand anything. How did the cow get this disease? How did this happen here?'"[121]

For some, the cost has been much higher. Consider the evidence taken by Lord Phillips at the beginning of the British BSE Inquiry concerning the circumstances of Clare Tomkins, who fell ill with vCJD in October 1996. Clare, an animal lover, had been a vegetarian since 1985, when she was thirteen. During the early part of her illness, incorrect diagnoses included depression, anorexia and agoraphobia. She underwent electroconvulsive therapy. Her father said: "The most harrowing thing was sometimes in bed at night...she howled like a sick injured animal. She looked at you as though you were the devil incarnate. She started to hallucinate."[122] At the time her father gave his testimony to the BSE Inquiry, she was still alive but bedridden, incontinent, blind and apparently unaware of her surroundings. She died a month later.

What happened to Clare? BSE and vCJD belong to the family of formerly recondite illnesses called Transmissible Spongiform Encephalopathies (TSEs). They are termed "transmissible" because they can be passed from one person or animal to another; they are "spongiform" because the brain becomes speckled with holes invisible to the unaided human eye; and they are called "encephalopathy" because the symptoms are caused by damage to the brain and nervous system. The version of TSE contracted by cattle is called Bovine Spongiform Encephalopathy (BSE), popularly known as "mad cow disease." TSEs are recondite no longer. Medical science now considers TSEs to be potentially a major human health risk, and a present global threat to commercial animal husbandry.

Until the onset of BSE, the sheep disease called scrapie was the only commonly known TSE. For the past 250 years, scrapie was a nuisance, but not a threat. Infected animals rub themselves constantly against fence posts to relieve their itchy skin, and eventually they die. But scientific studies of scrapie over the last century have shown that the illness is quite distinct from other infectious diseases of sheep. In fact, only some rare human diseases, notably Creutzfeldt–Jakob disease, are similar to it.

The incubation period of TSEs is long, spanning decades. The infectious agents survive physical and chemical environments that kill other life forms. Extremes of heat and radiation and powerful chemicals such as formaldehyde cannot kill TSEs. The infectious agent is also elusive. It does not seem to be a conventional infectious particle such as a bacterium or virus.

Scientific opinion on the etiology of TSEs has achieved a consensus on some important issues. There is general agreement on the importance of a single molecule, the prion protein. This protein is found in all warm-blooded animals from chickens to humans. However, it is not necessarily an essential protein. Transgenic engineering has constructed mice without prions, and they seem to enjoy happy and healthy lives. The prion is a membrane protein found on the surface of many different cell types in the body. It can be dissolved in chemicals and digested by enzymes. But in TSE, the prion changes dramatically. It cannot be dissolved or digested by enzymes. It changes its shape and begins to build up in cells. Scientists believe that this buildup explains why infected cells die. The death of cells causes the loss of brain functions and the appearance of the microscopic holes in the brain. The conclusion that the prion plays a crucial role in TSEs was based on the fact that the genetically engineered mice that lack it cannot be infected with agents such as scrapie that kill normal mice.

But how was the disease being spread? Not long after BSE was discovered, outstanding epidemiological work identified the central role played in its spread by the bone meal fed to cattle. Researchers hypothesized that the first cases were found in cows that ate ground-up sheep infected with scrapie. In other words, it was thought that BSE was scrapie in cattle. At the time, this offered reassurance to the British public who had been eating infected sheep for at least two hundred years and had suffered no apparent ill effects. However, there was no evidence of a link between scrapie and classic CJD.

In 1988, faced with a mounting BSE crisis, the British government set up a working party "to advise on the implications of Bovine Spongiform Encephalopathy and matters relating thereto,"[123] headed by Sir Richard Southwood, professor of zoology at Oxford University. Based on the evidence available at the time, it was possible for the Southwood Report issued in 1989 to tell the public that the risk of transmission of BSE to humans "appeared remote." It is only fair to say that the Report contained caveats: "Our deliberations have been limited by the paucity of the available evidence. Further research in this area is essential" and "if our assessment of these likelihoods is incorrect, the implications would be extremely serious."[124] The Report made two recommendations that it considered precautionary, that sick cows be taken out of the food chain and that bovine offal not be used in baby food.

Later, the BSE Inquiry asked a realistic question: if it was reasonably practical to take precautions with regard to baby food, why was it not also reasonable to take precautions with respect to adult food, especially in view of the working party's own statement of the possibility that their assessment might be incorrect?

It is difficult to exaggerate the importance of the Southwood Report. Right up to 1996, it was cited as if its conclusions, minus the caveats, were scientifically certain, rather than a statement of provisional opinion. In fact, the scrapie hypothesis was not rigorously tested until 1997 by feeding cows with material from scrapie-infected sheep.[125]

The British civil service practised secrecy and self-censorship. For example, in 1987, a member of the State Veterinary Service tried to publish his findings on BSE. The Assistant Chief Veterinary Officer, who was his superior, wrote to him:

I am now confirming that the letter to the Veterinary Record which I cleared earlier in the week should not be published. I explained to you that this condition had been discussed by the CVO and the Director of the CVL [the Central Veterinary Laboratory], and because of possible effects on exports and the political implications it has been decided that, at this stage, no account should be published.[126]

In his penetrating analysis of the findings in the BSE Inquiry, published in the *London Review of Books*, Hugh Pennington commented on this and other evidence of an official "cover-up": "We should not be surprised to find secrecy in the civil service, but there is something especially bad about secrecy and censorship in science. They are completely at odds with its norms: universalism, organised scepticism, communality, humility and disinterestedness."[127] The chair of the BSE Inquiry, Lord Phillips, summed up the public harm done by the tragedy, not merely to the reputation of scientists such as those who wrote the Southwood Report, not only to bureaucrats of the British government, but to the trust of ordinary people in Britain and Europe in the safety of their food. "When on 20 March 1996 it was announced that cases of new variant CJD were probably attributable to contact with BSE before precautionary regulations were introduced, the reaction of the public was that they had been misled, and deliberately misled, by the Government."[128]

5. The Precautionary Principle

The BSE crisis was only the worst of a number of food-related incidents in Europe,[129] which have led the public and government regulators to insist on what is known as the Precautionary Principle with regard to bioengineered food. The European regulatory mechanism has been developed to manage environmental or health risks arising from incomplete scientific knowledge of a proposed activity's or technology's impact. In other words, when the stakes are high due to scientific uncertainty or lack of knowledge, "it is better to err on the side of protecting human and environmental safety than to err on the side of the risks: 'better safe than sorry.'"[130] In the 1970s, the Precautionary Principle entered European environmental politics. Since then, it has become one of the principal tenets of international environmental law, appearing in more than 20 international environmental laws, treaties, protocols and declarations.[131]

A practical instance of its application occurred in 1999 when Canadian canola exported to Europe contained genetically modified seeds, despite official documentation to the contrary. Adventa Canada sold the seeds to four countries in Europe –

Sweden, Germany, France and Britain – where they were planted in the spring of 2000. A spokesperson for the company said that the company had no way of assuring that the wind or bees don't spread genetically modified traits to crops of conventional canola seeds. "The only way to make sure its conventional canola seeds don't get mixed up with GM varieties is to grow them in a country where GM canola is banned."[132]

This and other instances of the application of the Precautionary Principle remain controversial, although it is invoked because of the issue of risk. Some, if not all, food that has undergone gene splicing will benefit human health. How do we provide protection against the production of new and dangerous allergens or against a repetition of the 1989 incident in Seattle when people died from taking a genetically engineered food supplement?[133] The benefits of herbicide-resistant transgenic plants must be balanced by the risk of these hardy strains invading other crops as "super weeds"[134] and the eventual adaptation and emergence of insects even tougher than the plants. Although the benefits are clear regarding the potential introduction of drought-resistant maize or salt-tolerant rice, still those who are knowledgable about the need for biodiversity warn us that the novel varieties introduced could harm the indigenous crops of these countries.

These are some of the predictable risks. What is of even greater concern is that these new and complex procedures known as transgenic modifications could produce unanticipated, disastrous outcomes. Monsanto has offered us the assurance that there is "no evidence" at the moment that "altering genes could lead to unforeseen problems."[135] This is another example of the logical fallacy referred to earlier as "begging the question." One of the most commonly cited answers to the sort of faulty logic in Monsanto's guarantee is that it does not respect the distinction between "absence of evidence" and "evidence of absence" when assessing and managing risks.[136]

6. The Vatican's Position on the Risk of GMOs

There is nothing like consensus or unanimity on the issue of risk. With their usual caution, Pope John Paul II and his advisers in the Pontifical Academy of Life have given their guarded approval to transgenic modification of food. The 1999 collection of articles published by the Academy and entitled *Biotecnologie Animali e Vegetali Nuove frontiere e nuove responsabilità* discusses many of these issues. Among the most difficult is whether the engineering of life simply for human benefit does or does not manifest a lack of humility or, using the words of Iris Murdoch, of "selfless respect for the independent existence" of the creatures and life forms that belong to the world in which we live.[137] The Vatican position seems to be that while it is impossible to remove all danger of damage and even disastrous outcomes, the risks involved in GM food are less than their harshest critics would claim and that their potential benefits, particularly for the poor, are greater than these same critics would argue. Tim Dyson, D. Gayle Johnson and other population experts make a case that it will become increasingly difficult, although not perhaps impossible, for the world to feed itself in the future without these foods.[138]

The position taken by the Vatican is rational, but it is not the only rationality that should be exercised in this question. It is reasonable to accept the audit of potential benefits and risks of GMOs done by sound science, but reject the benefits and ban the whole technology on the basis of a different risk assessment. There is no rational procedure available to settle the question of a proper trade-off between risks and benefits, between "playing safe" and "reaching for the sky." Catholic tradition has spent considerable intellectual and moral resources on how we can resolve moral uncertainty. Belonging to the general discipline of casuistry, this topic has historically been called "probabilism."

7. Probabilism

In applying the moral theory of "probabilism" to the question of the risk involved in gene splicing, one becomes aware of the burden of history. Today, in meetings convened by the agriculture ministers of the European Union, moral decisions are taken around risk management in this question. Three hundred and fifty years ago, in the darkened hush of a baroque church, a whispered discussion might take place between a penitent and his or her confessor in the curtained confinement of a confessional box. This was the context within which probabilism has developed.

The resolution of debates surrounding probabilism informs the Catholic imagination in various ways. The papal teaching office condemned extreme rigorism, a demand for certitude in order to take a conscientious action, just as much as laxism with regard to one's free choice to take a moral risk. What this means is that the typical Catholic will try to follow a middle path in our question between "playing safe" and "reaching for the sky" with regard to using the products of GMO technology.

Probabilism comes from the Latin *probabilis,* meaning "provable." It has to do with any belief that is reasonable without being certain. In terms of moral systems, it means that when conscience is in doubt about the morality of a particular course of action, "if an opinion is probable, it is licit to follow it, even though the opposite opinion is more probable." This is the formulation in a 1577 publication of Bartolomeo Medina (1528–80), a Dominican professor of theology at Salamanca, Spain, who was known to the mining industry for his 1557 invention of the patio process technique to extract silver from ore using mercury. Medina is even more important in the history of moral theology thanks to his thesis that "a doubtful law does not bind." According to Medina, the authority that would adjudicate whether a decision was probable, or tenable, or not, would be "wise men with excellent arguments."

Medina's formulation came out of a long medieval tradition of teaching the faithful about the kind of knowledge required for moral judgment. While theologians consistently taught that it was sinful to act while in doubt, earlier theologians generally taught that, should doubt persist, one should take the safest course in terms of action that would most likely avoid sin and lead one to salvation.[139] If one were to relieve Medina's formulation of its arcane wording and relate it to the question of the moral use of GMOs, the thesis of probabilism says that a person who deliberates about whether or not he or she is obliged by some moral, civil or ecclesiastical law to refrain from their use may take advantage of any doubt whether any law obliges him/her to abstain.

The situation in life out of which the theory of probabilism emerged was the confessional. Penitents in the sixteenth and seventeenth centuries often approached priests exercising this ministry for advice and absolution from their sins in the sacrament of penance. Their moral life was a complex reality consisting in many laws: eternal laws of God, canon law of the Church, civil law of their place of residence. The confessor had to decide whether or not to oblige his penitent when one or other of these laws seemed to him to be less than absolutely applicable and binding in his or her case. This would be apparent to the confessor, to other confessors previously consulted, and to the penitent. The penitent might say in good faith that he or she believed that the law in question did not apply to him or her with sound reason based, perhaps, on the opinion of another confessor. The question behind Medina's thesis is whether the confessor is morally bound to judge the penitent by his own opinion, which, in his view, is more rational. He may also accept the apparently less sound, though still reasonable, opinion of the penitent or the other confessor. Medina's thesis, the basis of probabilism, would answer in the affirmative.

At the same time, Jesuit theologians developed the option for liberty from the law involved in probabilism. Gabriel

Vasquez, SJ (1551–1604) separated the two components of this moral theory, its *intrinsic* probability based on "excellent arguments" and its *extrinsic* probability founded on the authority of wise men. In the long run, this distinction became problematic because it ended with a situation in which moral probability became more a matter of counting how many and how prestigious one's "wise men" were than in examining carefully the reasons behind the position being taken.[140]

Debate over the way this theory could and should work continued to take place. Many distinguished theologians opposed the Jesuits and their moral theory. In the process, certain distinctions that may be helpful for the GMO question of risk were articulated. For instance, Prosper Fagnanus (1588–1678), a "rigorist" and therefore a critic of the Jesuit theory, distinguished between "probable certitude" and "certitude arising from probabilities." Despite the fact that St. Thomas Aquinas had used the first concept, Fagnanus considered it to be an oxymoron. According to philosophers Albert Jonsen and Stephen Toulmin, in his view, the latter expression ("certitude arising from probabilities")

> can be properly applied to certitude, an "assent of the mind," engendered by the convergence of many probable arguments toward the same conclusion. This, says Fagnanus, is a genuine certitude, even though of a different kind from the certitude reached by formal demonstrative reasoning; and such a convergence of probabilities toward a certain conclusion can be a rule for action.[141]

Within the context of real and potential risks involved with gene splicing, there is a "certitude arising from probabilities" that, as time passes, the process itself can be deemed safe. Two key questions arise from this certitude: whether it is begging the question by confusing "absence of evidence" with "evidence of absence" of risk and whether GMO technology is just.

In 1679, the official teaching of the Catholic community, through a decree promulgated by Pope Innocent XI, rejected the approach known as laxism. According to this approach, a person might consistently adopt the most permissive and least well-grounded opinion on a moral question such as the development and use of gene-splicing technology.[142] The following year, the Holy See allowed two widely supported views on probabilism to stand without condemnation, but it did not embrace definitively one or the other. According to the first, probabilism, one could follow a solidly probable opinion; according to the second, *probiliorism* (i.e., "more probable"), only a more restrictive alternative might safely be followed. Another view, called *tutiorism* or rigorism, held that the strictest opinion was the only safe one to follow. *Tutiorism* is roughly equivalent to the Precautionary Principle, previously discussed above. As it would be used in moral theology, the Holy See rejected *tutiorism* in 1690.[143]

Although these antique terms are not in common use any longer in the moral teaching of the Catholic community, the approach taken in this casuistry is still in force in some difficult moral issues.[144] With respect to the development and use of GMOs, the Vatican position would seem to be *probiliorism*, inasmuch as it is clear that the freedom now allowed in the field by the official teachers of the Church, namely its bishops in union with the pope, probably would be rescinded if credible scientific evidence emerged that the technology itself was seriously risky to humans or other living creatures. Many churches and ecumenical organizations, including their umbrella group, the World Council of Churches (WCC), have chosen to take a different approach. They are concerned about the dangerous consequences of the global proliferation of GMOs.[145] Whereas the Vatican's methodology to date has been one to be prepared to react to any evidence of risks, the WCC anticipates serious problems and urges care and caution now.

This relates to public policy. With respect to individual moral choices it would seem that there is no alternative except to insist on labelling transgenically modified food products so that consumers can make informed decisions as to whether or not to use them.

The GMO revolution can be seen as a crisis in the Chinese meaning of that word. In written Chinese, the word for "crisis" consists in two different ideograms, one superimposed on the other. The first is the word for "danger." The discussion on the risks involved with GMO food certainly falls under the rubric of this word. But "danger" alone does not constitute a crisis. On top of it one must draw the word for "opportunity." There is no crisis without risk; there is no crisis without opportunity.

Chapter 4

Transgenic Modification and the Eucharist

The Catholic moral preference for a mediating position between laxism and rigorism with respect to the question of the risks involved with GMOs suggests that we revisit one of the seminal works of twentieth-century theology, David Tracy's *The Analogical Imagination*. According to Tracy, underlying and providing a bulwark for secondary manifestations of Christian theology such as the theory of probabilism, there exist "conceptual languages."[146] In the history of Western Christian thought, two quite distinctive language traditions exist. Analysis of these two traditions can provide an explanation for the difference in approach to and evaluation of the ethics of GMO research and production taken by the Vatican and the World Council of Churches. Tracy calls the Roman Catholic language "analogical" and says that it is "a language of ordered relationships articulating similarity-in-difference." In this conceptual system, the Incarnation is the prime analogue used to interpret the whole of reality. Because of this incarnational focal point "the entire world, the ordinary in all its variety, is now theologically envisioned as sacrament – a sacrament emanating from Jesus Christ as the paradigmatic sacrament of God, the paradigmatic clue to humanity and nature itself."[147] There are distinctions and differences between God and the world, but the emphasis in Catholic tradition remains "analogies in difference."

Over against this sacramentally based language system, there is a competing framework which Tracy calls "dialectical language." Proponents of this theological construction such as Martin Luther, Karl Barth and Rudolph Bultmann would insist on the role of fundamental negation in distinguishing the Holy from human science in all authentically Christian language. This language system sees a rupture between God's revelation of salvation and the human condition, "a rupture at the heart of human pretension, guilt and sin – a rupture disclosed in the absolute paradox of Jesus Christ proclaimed in the judging, negating and releasing word."[148]

It is the thesis of this book that there is a distinctively Catholic perspective on the bioengineering of food. Those committed to the second language system described above, who are gifted with a "dialectical imagination," will find what follows in this chapter seriously erroneous. This exercise of the "analogical imagination" with respect to the virtualities that the new technology offers for re-expression of the traditional Catholic doctrines of the Eucharist is congruent, however, with Catholic perspectives on transgenic modification of bread and wine.

Two seemingly unconnected news items provided the stimulus for this final chapter on Catholic perspectives on genetically modified food. The first was the controversial decision by the Canadian Food Inspection Agency to allow the testing by Monsanto and Syngenta of genetically modified wheat. The test sites in five different provinces are secret[149] because of concerns over possible protests. Out of fear over potential market losses, representatives of 210 industry associations led by the Canadian Wheat Board sent a letter of protest over this decision to the Canadian Prime Minister, Jean Chrétien.[150]

A few days earlier, the U.S. bishops issued a statement in the form of a catechetical primer on the Catholic understanding on the Eucharist entitled "The Real Presence of Jesus Christ in the Sacrament of the Eucharist: Basic Questions and Answers." Following the question-and-answer format of medieval theologians such as St. Thomas Aquinas, the bishops answered in traditional ways questions like "When the bread and wine become the body and blood of Christ, why do they still look and taste like bread and wine?"[151]

The reason the bishops issued this document was a felt need. They had come to know that practising Catholics who regularly receive the Eucharist in their churches had, in their view, an incorrect or at least an incomplete understanding of who or what they are receiving. Especially in the United States, but in most other jurisdictions as well, Catholics think in a modern and scientific way, and their faith understanding tries to be congruent with this mindset. When discordance arises between their received faith doctrines and their contemporary way of thinking and speaking, the risk is that they will reject or modify the teaching in order to make sense of it for themselves and others. Now that field tests have begun on genetically modified wheat, it might be timely to consider the possibility of alternative forms of discourse, in addition to the Thomistic–Aristotelian categories used in their instruction by the U.S. bishops. The science of molecular biology offers us certain ways of speaking about the transformation of bread that could prove useful in our own discussions of eucharistic theology.

Speculative theology has always played a vital role in Catholic life. While not authoritative in the same way as the teaching of the pope, the bishops and their official collaborators, this form of theology has its own importance. For instance, it can provide believers with new and fruitful ideas about their faith. As well, it can offer explanations for those interested in what believers think and why they think it. Many contemporary theologians dedicate themselves to the need to re-express the mystery of Christ for their own age. For each historical age and in each culture, the mystery of Christ Jesus, the life of God's grace and favour made possible for humans to enjoy, and the various ways in which this life is communicated in the sacraments needs to be re-expressed, reformulated and pondered again. With the importance of scientific ways of thinking and speaking in our world today, what theologians say and how they say it may need some updating. Speculative theology today needs to engage itself with science because science is the primary way that people of our age, whether believers or not, think about what really matters.[152]

What follows is not a dialogue between the science of molecular genetics and eucharistic theology. Dialogue is an exchange of ideas between theologians and scientists, each from within their own field of expertise, on a common topic of interest, such as the definition of life.[153] Rather, the following discussion is a discourse. In a discourse, one party, in this case Roman Catholic theologians, examines the method and the way that bioengineering raises questions and solves problems, in order to speculate on the Eucharist.

Catholics who engage in the enterprise of theological discourse need to have a correct understanding of two related concepts. The first is *analogy*. In the discourse that follows, all talk about God and the things of God is analogous. The reason is that God is, in the words of the late and great theologian Karl Rahner, "holy mystery," whom we can approach only in an "asymptotic" way.[154] Using the mathematical theory that as curved lines approach infinity, they continually approach each other but never touch, Rahner created a metaphor or analogy for the simultaneous knowability and unknowability of God and the fact that total union with God's self is not possible within this world or after one's death. The second thing that we must keep in mind is that the relationship between faith and reason means that, combined, they enable us to understand the most important things in life better than either would on its own. Prompted by the official teaching of the Church at the First Vatican Council and by an encyclical of Pope John Paul II,[155] what follows is a reflection on various questions about the Eucharist within the parameters of theology that makes use of certain analogies taken from the scientific discourse used in discussing the genetic alteration of food. Borrowing can go in both directions, with molecular biologists discussing "substantial equivalence" in the same way that Thomas Aquinas borrowed Aristotelian philosophy, and, in the case of this present discussion, with a Catholic theologian using possibilities opened up by the decoding of the DNA molecule to understand the bread and wine that is converted in the Eucharist into the Body and Blood of Christ.

In this discussion, the general topic of "Catholic perspectives on the bioengineering of food" is served by the suggestion that an understated and subliminal aspect of the Catholic imagination, belief in the real presence of Jesus Christ in the Eucharist, makes it perhaps easier for Catholics to accept the process of genetic transformation than for those who do not believe in this theological teaching. In a sense, Roman Catholic Christians claim to have been part of this process for a very long time.

1. A Thirteenth-Century Example

An example of medieval theological discourse employing science to aid in understanding of the Eucharist would be Questions 75 to 78 in the Third Part of St. Thomas Aquinas' *Summa Theologiae*.[156] Aquinas' purpose was to answer certain questions about how the bread and wine become the Body and Blood of Christ; about the way that Christ is present in the Eucharist so that, for example, the accusation that Christians are engaged in cannibalism could be suitably answered; about the relationship between the visible dimensions of the bread and the Body of Christ; and about the role of language in the conversion. Studying the best science of his day, namely the work of Aristotle, then available to theologians for the first time in translation, Aquinas was able to solve several of these issues to the satisfaction of his culture and age.

Aristotle approached many of the philosophical questions with which he dealt by using the notion of substance. Thanks to this category, Aquinas was able to negotiate his way out of the problems that traditional understanding of the Eucharist presented to him. In reading the texts of Aquinas in their original language, one is struck by the fact that he rarely employed the best-known term used by subsequent ages, namely "transubstantiation." Thomas preferred to speak about *conversio*, "conversion," "changing into" rather than the neologism "transubstantiation" coined by his predecessor, Orlando Bandinelli (1105–81, who became Pope Alexander III). The consecration, Aquinas commented, resembles a natural change, yet because it is a special conversion, it may properly be called transubstantiation (3.q.75.4). His preferred term, however, was a scientific one used at the time, namely conversion, and this suggests that Thomas was engaged in a discourse with natural science about the Eucharist.

Consider, for example, the fact that we cannot see with our eyes the Body of Christ after the consecration. How can this be explained? Thomas found an easy answer in Aristotle with the notion of substance. The substance of any being, including material substance, cannot be seen by our ordinary sight but only by the "spiritual eyes" of our intellect (3.q.76a.7). For Aristotle and for Thomas, the substance, or the essence of what a thing really is, *quod quid est*, can be grasped only by our mind. The notion of substance, understood not as an object of our intellect only (as it was later in theology for Vasquez [1549–1604] and Suarez [1548–1617]) but as a reality, is foreign to modern science. However, for scholastic philosophy, substance was an ontological reality, a form of being (*ens in se*), one that is distinct from the way of being in an "accident" (*ens in alio*). Only God is a being *ens in se*.

Theology must remain within its own parameters when engaging in discourse using science. Thomas Aquinas was always careful to respect the demarcation lines. For instance, he wrote that the conversion of bread into Christ's body has a certain resemblance to creation and to natural changes, but, at the same time, it differs from both. What all three share is that something quite different emerges in the second state, as compared to the first. In creation, being follows upon non-being; in natural change fire, for example, follows air; in the sacrament of the Eucharist, the body of Christ takes the place of the substance of bread. Still, the consecration is more like a natural change than it is like creation, because it does not emerge from nothing. As well, in both a natural change and in a eucharistic consecration, something remains. In the natural change, it is the subject that loses one substantial form to take on another. In a eucharistic conversion, while it is true that the substance of the bread and wine do not remain, there is still the appearance of bread and wine. The consecration, therefore, is not a creation but a conversion because the "accidents" of bread and wine remain. Using the medieval hermeneutical device of "*synecdoche,*" that is, naming the part for the whole, we can say that the bread and wine "become" the body and blood of Jesus Christ. Expressed differently, the Body and Blood of Christ are contained under the same appearance formerly employed by the substance of the bread and wine (3.q.76.a.8).

In his own age, Aquinas' use of Aristotle's scientific discourse in his theological synthesis provoked controversy because of the fact that Aristotle was not Christian. As a method of communication, it has proven to be so successful that contemporary Catholics conversant with Thomism are often unaware of the scientific provenance of some of the words and concepts in which his theology was expressed. In the spirit of Aquinas' efforts to offer greater clarity on the doctrine of the Eucharist, let us examine some key ideas from science in order that, in their light, we may penetrate further the mystery of faith.[157]

2. How Biotech Genes Convert Organisms

One way to describe the eucharistic conversion is in terms of "new information." This way was used by Augustine and Aquinas and is congruent with modern ways of thinking in terms of new information.[158] In some mysterious way, the bread and wine receive from God through the Risen Christ within the Church certain information that leads to a transformation into the Body and Blood of Christ. Catholic theological traditions have used the concept of the "Word of God" to convey the "new information" that catalyzes this conversion.

Information is crucial to genetics as well. In the century that followed the death of Gregor Mendel, the science he pioneered has taken various directions. In 1900, three biologists working independently rediscovered Mendel's laws, which stated that the characteristics of organisms are determined by units of heredity. William Bateson introduced the term "genetics" in 1906, and Wilhelm Johannsen first used the word "gene," intended to denote the hereditary units themselves, in 1909. By 1930, Thomas Hunt Morgan and his colleagues, working with the fruit fly *Drosophila,* had demonstrated that genes are arranged in a linear fashion within chromosomes.[159]

With the discovery and decoding of the DNA molecule the science of genetics has been called by some "genomics," after the genome or complete map of DNA. Its rise in the last century paralleled the science of computing. This chronology is no accident – both sciences concern the storage and timely retrieval of information.

Electronic computers, the first of which appeared in the 1940s, use the simplest of mathematical arrangements to encode their information. Everything known by a computer's database is broken down into a string of ones and zeros known as binary digits or "bits." These bits are then combined into strings, eight binary digits long, known as "bytes." "By employing these bits and bytes in combination with memory stores and electronic 'gates' that perform mathematical operations on them, it is possible to build a device that can, at least in principle, solve any problem that can be written down as a finite series of logical steps."[160]

The development of computers is impressive, but not nearly as impressive as the contemporary human discoveries about how cybertechnology echoes the way God has put living things together. In a sense, life is also, essentially, a process of digital computing.

In the case of living things, the code is quaternary, rather than binary. DNA, or deoxyribonucleic acid, is an abbreviation that our computer spellcheckers now accept as a word in its own right. Often described as the blueprint for life, DNA is a threadlike molecule that, as we have seen, was first described by Rosalind Franklin at King's College in London, and then by James Watson and Francis Crick in Cambridge fifty years ago. In fact, DNA is not a blueprint in the sense of a sketch or diagram, but rather a physical entity that can and sometimes does break when, for example, it is forced through the narrow end of a syringe.

DNA consists of four small chemical bases: thymine, guanine, adenine and cytosine, abbreviated T, G, A and C, respectively. They are linked together in long chains supported by a chemical "backbone": the strands forming the famous double helix, first portrayed by Watson and Crick, about which more will be said later.[161] DNA is ubiquitous in life because it is contained in almost every cell.[162]

RNA (ribonucleic acid), closely related to DNA, is also present in every cell of every organism and is more active than DNA. RNA differs from DNA because it has one different base, uracil (U), which replaces thymine (T), and because it is usually single-stranded rather than double-stranded.

DNA has one job and one job alone: to carry genetic information. RNA carries these genetic records, but it also serves as the "genetic material" in most viruses, a task performed by DNA in all higher organisms.

A gene is not, therefore, a "thing," and not a written plan. Unlike a gene, DNA is a physical entity. But the blueprint analogy does not work for it, just as it did not for a gene. The analogy is faulty in the case of DNA, because it is not one configuration but rather a set of instructions, similar to the pre-Vatican II textbooks in moral theology which contained all possible answers to every ethical dilemma, as well as a clear description of every sin and its particular malice. The chapters in the DNA "textbook" are laid end to end, as in a long scroll. University of Saskatchewan scientist Alan McHughen compares DNA to a recipe book. "A gene is not a physical entity the way DNA is; rather it is a unit of biological information ... think of a gene as one of the recipes contained in the DNA. The entire complement of genetic information of an organism is the genome. Think of it as an encyclopaedia of genetic recipes."[163]

While the binary digits in computers are electronic, the quaternary digits in DNA are chemical. Having a quaternary code means that DNA can put together shorter combinations than the bytes of a binary system. These combinations, known as codons, are three bases long, meaning that there are 64 possible combinations that provide the organism with the information it needs; in the case of the human body and other organisms, this information concerns the making of proteins. In other words, in some sense codons cause the body or, more accurately, the amino acids in the body to produce proteins.

Proteins are the workhorses of biology. Almost every molecule in the body is either a protein or the result of a protein's activity. Proteins are made of smaller molecules, known as amino acids, strung together in chains that are usually several hundred amino-acid units long. Biology employs 20 different sorts of amino acids in the construction of proteins, and most of the 64 codons correspond in meaning to one (and only one) of these amino acids.[164]

In the 1953 announcement to the world of their discovery that DNA is genetic material, published in the journal *Nature*, Watson and Crick concluded their article by saying something that seemed innocuous at the time but which has proven to be very important: "It has not escaped our notice that the specific pairing we have postulated immediately suggest a possible copying mechanism for the genetic material."[165]

DNA is a two-stranded molecule in which the strands twist around one another in a specific pattern known as the double helix. If they were untwisted, they would look like a ladder. The two uprights of this ladder are molecules of sugar called deoxyribose. The rungs are each made of two bases. Watson and Crick pointed out that the genetic material lends itself to copying because of the fact that, although the individual bases are of different sizes, their usual pairings create units of identical shape and size – A pairing with T and C with G. Since the rungs fit between the uprights in each direction, the bases attached to any given upright can come in any order. However, the order of the bases on one upright necessarily specifies their order on the other. "That means that if a DNA molecule is unzipped up the middle, each half of it can be used as a template to recreate a whole molecule identical to the original. This replication happens every time a cell divides. It is the core of life's operating system."[166] Dr. Paul Berg of Stanford University in California led a team that used this potential in the DNA molecule and some restriction enzymes to cut two DNA molecules from two different sources. The hybrid DNA molecule they created by splicing these two separate pieces together is often referred to as recombinant DNA. Genetically modified (engineered) organisms are made of cells that contain one or more recombinant DNA molecules.[167]

There are four core axioms in the discourse of genetics, which are fundamental to molecular biological discourse today: 1. All organisms consist of cells and cell products. Some, like bacteria, have only one; most organisms have many, but these

cells are not quite large enough to be seen by the naked eye.
2. Each cell in an organism contains the same set of genes that
we call the "genome."
3. The genome contains all the genetic information needed to
make the entire organism. Theoretically, we could take a mature cell from, say, the liver of a mongoose, and induce it to
grow an entirely new mongoose, a genetically identical clone
of the parent mongoose that originally donated the liver cell.
4. All organisms share the same genetic language, known as
the language of DNA. A gene from a mongoose can be read and
properly understood in a human or in a plant cell.

Within the context of the new biotechnology of food grains,
genes convert organisms by conveying to them particular pieces
or units of information.[168] With information, something that is
essentially intellectual, we can use genes to change something
that is essentially physical. Robert Shapiro, the former CEO of
Monsanto, has described biotechnology, which puts to human
use, whether good or not, the discoveries we have made about
the genome in the following way: "Biotech is a subset of information technology. It's a way of encoding information in nucleic acids as opposed to encoding it in charged silicon. I put a
gene, which is information, into a cottonseed, and I don't have
to spray stuff on the crop in order to control insects...."[169]

3. Transcendental Implications of the Debate over Food Safety

The politicians, scientists, administrators and technicians
who are entrusted with the regulation of the food supply in
any country and with the novel plants and animals that have
emerged with the age of biotechnology carry serious fiduciary
responsibility. The phrase "fiduciary responsibility" refers to the
fact that some members of our community are entrusted to
provide care and oversight for others in such matters as health

and safety. Critics claim that, faced with the unprecedented developments in biotechnology, these members of society have succumbed to the blandishments of powerful corporations to rush products through the process of study and testing, so that they can be approved for human or animal use.

Consider, for example, the Taco Bell™ controversy that erupted in 2000. Bioengineered corn, not approved for human consumption, was found in taco shells and other food sold in the U.S.A. Developed and marketed by Aventis under the trade name of StarLink™, this version of Bt (Bacillus thuringiensis) corn contains the transgenically introduced protein Cry9C, one of 43 proteins in the bacterium Bt, each of which is toxic to different insects. The Environmental Protection Agency (EPA) of the United States was willing to approve Cry9C in 1998 only for animal feed because it could not rule out the possibility of allergic reactions in humans. In September 2000, a group known as the Genetically Engineered Food Alert Campaign found that taco shells manufactured in Mexico for Taco Bell restaurants and distributed by Kraft contained StarLink™ corn. StarLink™ next showed up in Safeway's brand of taco shells, which led managers to pull them from their shelves. In mid-October, Mission Foods, the largest U.S. manufacturer of tortilla products, recalled its tortillas, taco shells and snack chips.[170] In the same week, Kellogg, the American food company, shut down one of its plants because of the same problem.[171] Although many people ate these food products and no allergic reactions were reported, the possibility could not be dismissed. Government regulators, not only in the United States but in Japan, Britain and Canada[172] among others, have expressed the concern that they simply cannot keep genetically modified and unmodified crops separate.

The owners of the patent for StarLink™ corn, Aventis Corporation, expressed appropriate alarm and took remedial action. On October 12, 2000, Aventis announced that they had cancelled their registration of StarLink™ corn, which meant that it could no longer be planted for agricultural purposes.[173] The subsequent recall of seed has cost Aventis more than $100 million, but these efforts at damage control were unsuccessful. Faced with mounting evidence that StarLink™ had entered the human food chain throughout the world (an event that gives new meaning to the phrase *fait accompli* because the accomplished facts have also been digested). Aventis then revised its strategy. It petitioned the EPA for a time-bound exemption from the original prohibition against the production and sale of StarLink™ for human consumption. By time-bound, they meant that the permission granted to them would be temporary. They argued that the risk of an allergic reaction to their version of Bt corn was remote. They further made the case that StarLink™ met the "reasonable certainty of no harm" safety standard of the Food Quality Protection Act and should be retroactively approved for human consumption.[174] The EPA, under serious criticism for its original conditional approval of StarLink™, had not responded at the time of writing.

The revised argument of Aventis would seem to be the following: In order to obtain a patent on Monsanto's original discovery of how to transgenically modify maize in order to produce Bt corn, which contains its own insecticide against the corn borer, Aventis has added the Bacillus thuringiensis (Bt) protein Cry9C. Their version of Bt corn is "substantially equivalent" to the originally produced type which, in turn, is "substantially equivalent" to maize that has not been transgenetically modified. Therefore, the EPA should approve it with the expectation that little or no harm would come to those who have consumed it or will do so in the future. The approval would be time-bound so that it could more easily be retracted, should there prove to be any harm to consumers.

The origin of the concept we call substantial equivalence[175] comes from the conventional breeding process. "Plant breeders work primarily with highly refined breeding lines whose genetic heritage is known, and whose progeny have been evaluated.... The expectation, borne out by years of successful crop variety development, is that 'barley is barley is barley'...."[176] Using a definition established originally in 1993 by the Organization for Economic Co-operation and Development (OECD), the Food Directorate of the Health Protection Branch of Health Canada would describe the notion of substantial equivalence as follows:

> [S]ubstantial equivalence embodies the idea that existing organisms used as food or as a source of food can be used as the basis of comparison when assessing the safety of the human consumption of a food or food component that has been modified or is new.
>
> If one considers a modified traditional food about which there is extensive knowledge on the range of possible toxicants, critical nutrients or other relevant characteristics, the new product can be compared with the old in simple ways. These ways can include, *inter alia*, appropriate traditionally performed analytical measurements or crop-specific markers, for comparative purposes. The situation becomes more complex as the new products lack similarity to old established products or, in fact, have no conventional counterparts.[177]

A year before the StarLink™ controversy, Eric Millstone and two colleagues published an article entitled "Beyond Substantial Equivalence" in *Nature*. It was criticized at the time as "ill-informed and ill-advised"[178] by two scientists working at an American university and by food industry spokespersons, in a parody of the title of the offending article, as "beyond reason."[179] In view of the StarLink™ developments the following year, their contribution has proven to be prescient. Despite the efforts of Aventis, the U.S. EPA has not (yet) said that StarLink™ is substantially equivalent to Bt corn, let alone to maize.

Millstone et al. introduce their remarks with the assertion that the concept of substantial equivalence itself has never been properly defined and is therefore vague. They add that it is "a pseudo scientific concept because it is a commercial and political judgement masquerading as if it were scientific. It is, moreover, inherently anti-scientific because it was created primarily to provide an excuse for not requiring biochemical or toxicological tests." In other words, they suggest that those in the biotechnology and food production industry who seek to avoid the expense and delays that would follow rigorous testing of their novel food products are using the concept for their own purposes. In a section of the article entitled "Trying to have it both ways," they criticize a 1990 United Nations/FAO report where the concept, but not the term "substantial equivalence," was first used, because the report "implies that GM foods are in some important respects novel, but it then argues that they are not really novel at all, just marginal extensions of traditional techniques."[180] The same temptation of wanting to "have it both ways" is faced by corporations who patent their new seeds and other bioengineered discoveries and then claim to regulators that they are substantially equivalent to the existing organism that has not been transgenically modified.

Food safety, under the rubric of the so-called precautionary principle, is the primary reason why the authors of the controversial *Nature* article argued that we should reject the concept of substantial equivalence. This concept is currently applied by government regulators throughout the world. The precautionary principle was discussed in detail in the last chapter, where it referred to the most cautious approach that can be taken by risk management.

Behind this practical issue lurks a metaphysical one. Is a genetically modified organism, whether altered using recombinant DNA technology or by conventional plant breeding methods, really the same as or substantially equivalent to what went before or have we created something entirely new? Has "transubstantiation," to use Thomas Aquinas' and the thirteenth-century's neologism, taken place?

It will be argued here that the answer is negative. The affirmative answer would have to make use of a paradigm that is not appropriate for genomics. In fact it is an example of an unacknowledged import of a Thomistic philosophical concept originally used for eucharistic theology.

Employing a crucial item of genetic information not found in the existing plant's genome but in that of a bacterium's genetic endowment, for example, we are able to create a new subspecies with a resistance to certain herbicides or a built-in and fatal toxicity for insect predators that would kill the unmodified subspecies. Perhaps the use of "species" in this context is making use of an old paradigm, one that has been created and encouraged by traditional Judeo-Christian discourse about creation and nature.

Until the decoding of the DNA molecule and the possibilities that it afforded us, the basic unit of nature was the species. Darwin argued that if we understand how species come into and go out of existence, we will then understand life itself. The Judeo-Christian tradition based its thinking on "speciesism" on the first two chapters of Genesis, the creation of the world by God, species by species, culminating in the climactic event of the creation of the first human persons. The biblical story of Noah depicts this ancestor saving our living world by herding members of each species into the Ark.

The ascendant paradigm today would try to claim that the essential unit of nature is not species but DNA. Those who find this formula reductionistic dispute this paradigm. But, understood correctly, it is a helpful shorthand to explain how, because of the genetic discoveries of the 1950s, farmers have been able to remove a relevant gene from the bacterium *Bacillus thuringiensis* (Bt) in order to produce a protein that is toxic to certain target insects such as butterflies and caterpillars. This is not a reductionistic use of DNA that violates nature because the reason this can happen is precisely that nature allows it, and because the modified and unmodified corn are not separate substances or species.

The fourth core axiom of the discourse of genetics is that "nature doesn't care in what a gene is located; all nature cares about is that it is a gene."[181] If one is committed to the notion of nature as a realm of distinct substances and species, taking a gene from the Bt bacterium and moving it into maize might seem to violate nature. If one is involved with modern biology, that is precisely what is natural. If it were not natural, the genetically modified corn would die and certainly would not be toxic to its traditional enemy, the corn borer. Thus, one implication of the debate over substantial equivalence is that we are no longer dealing with substances or species, but with genetic manipulation and modifications of genes and their DNA for the benefit of humans and animals.

Could it be possible that the Aristotelian scientific discourse used by Thomas Aquinas in his theology of the Eucharist concerning conversion of a substance has influenced the contemporary arguments about whether or not transgenically modified agricultural products are substantially equivalent to the unmodified species? The position taken by Millstone, Brunner and Meyer implies that these GMO products are not substantially equivalent. In other words, just as the bread and wine used in the Eucharist are not substantially equivalent to the Body and Blood of Christ after the solemn prayer of the Church at Mass, so a GMO product is not the same as the original substance. In neither case should the result of this transformation be treated in the same way. This is certainly the position of the Catholic community who reverently reserve in the tabernacle the unused hosts after Mass.[182] It is also why Millstone and his colleagues would advocate that we undertake careful tests and studies of the GMO products we have produced and not treat them in the same way as their unmodified versions.

Those who agree with Millstone et al. on the need for more stringent regulation of GMO food would probably disagree with them on these further implications. They would do so because we have illegitimately imported a theological discourse around substance into the field of genetics where the discourse is controlled by the discoveries of ways that genes convey information. If the Thomistic notion of transubstantiation is not really applicable to biotechnology, one could also ask whether the discourse of modern genetics could or should be imported into our efforts today to understand better the mystery of the transformation of bread and wine into the Body and Blood of Christ.

4. The Eucharist as Divine Transformation of Bread and Wine

Catholics understand the Eucharist as a special way to share in the new union of humanity with God brought about by the salvific activity of Christ, which St. Paul calls a new creation (Romans 6:4; Galatians 2:20). In the Eucharist, St. Paul finds a unique expression, not only for union of Christians with Christ, but also of Christians among themselves (1 Corinthians 6:15-17). For Paul, the Eucharist is, above all, the Lord's Supper, at which the new people of God eats its spiritual food and consumes its spiritual drink (1 Corinthians 10:3-4).

Since the end of the first century CE, Christians have struggled with the meaning of Jesus' actions and words at the Last Supper before he died on the cross. Should his words "This is my Body" (Luke 22:19) and "This is my Blood" (Matthew 26:27; Mark 14:24), pronounced over the bread and wine at the meal with his apostles, whom he then commanded to "do this in remembrance of me," be taken literally or not, and, if so, how should they be understood in later centuries? A variety of solutions have been proposed over the past two thousand years.

The Jerusalem Catecheses instructs neophyte Christians about the Eucharist in the following way:

... Therefore, it is with complete assurance that we receive the bread and wine as the body and blood of Christ.... Do not, then, regard the eucharistic elements as ordinary bread and wine: they are in fact the body and blood of the Lord, as he himself has declared.... You have been taught and you are firmly convinced that what looks and tastes like bread and wine is not bread and wine but the body and the blood of Christ....[183]

Later in theological history, Berengar of Tours (d. 1088) reacted against the extreme realism of this theology. He argued for a more symbolic understanding of the sacramental presence of Christ in the Eucharist. However, in doing so, Berengar found himself facing ecclesiastical problems. Signs were understood at the time to be merely pointers to something else; in themselves, they did not participate in what they signified. Berengar said that the bread and wine "are not the true body, nor the true blood, but a figure or likeness...."[184]

Berengar's important contribution to the theology of the Eucharist was his careful analysis of the process of change whereby the elements of bread and wine become the sacrament of the Body and Blood of Christ. His conclusion was that the "substance of bread does not change into the sacrament of the body of the Lord."[185] During the theological age known as the Carolingian period, the question had been asked: What are bread and wine? Within the context of Eucharistic theology, however, the real question, which Berengar asked and answered, was: What change takes place and how is it to be conceived? His answer to the first question, as to what bread and wine were, is that they are the sum of their sensible properties. As to what change takes place, Berengar said that the sensible properties of the bread and wine remain unchanged; nothing happens to the substance of the bread and wine, to their *veritas*, but they do become symbols through which Christ works spiritually in the hearts of those with faith.

The official reaction to his ideas was negative. Pope Leo IX, together with his fellow bishops, undertook an inquiry. Ultimately, a synod was convened in Rome in 1059. It included their most important theologian, Lanfranc of Bec (1010–89). At the end of the synod, Berengar had to burn his books before Pope Nicholas II, Leo IX's successor, as well as articulate a formula that upheld the contemporary understanding of realism at the Eucharist.

This was, by no means, the last controversy over what exactly constituted the change in the bread and the wine in the Eucharist. It was a key point of contention between Martin Luther and the papacy in the sixteenth century, and between him and his fellow reformer Carlstadt. As opposed to the Thomistic theory of transubstantiation, which was the position of most theologians of the time, Luther held for "consubstantiation" according to which Christ is "in, with, and under"[186] the bread and wine in much the same way that the divinity was (and is) united to the human nature of Jesus of Nazareth in the so-called hypostatic union.

Both Berengar's and Luther's views of the Eucharist were rejected by official Catholic theology during the sixteenth century Council of Trent, which affirmed the notion of conversion of the bread and wine into the Body and Blood of Christ.[187] But Trent did not say that one had to employ the philosophical theory of transubstantiation in order to explain this conversion.

Catholic theologians today are pleased about this because they are, on the whole, unhappy with the transubstantiation paradigm. They feel that it can lead to a static notion of the real presence of Christ in the Eucharist. By this, they seem to mean that it suggests a "vertical supernaturalism"[188] which would make the distinction between nature and grace too radical.[189] At present, however, they have not arrived at a commonly accepted approach to a theology of conversion to replace the theory of transubstantiation.[190]

A static notion of the eucharistic presence of Christ in the bread and wine, one that is extrinsic or incidental to the nature of the bread and wine itself, is similar to the problem of reductionism in biology. Reductionism, within our present discussion, could be defined as the insistence that molecular genetics is best served by arriving at simple conclusions rather than by adding levels of complexity that make definitive conclusions difficult to achieve.

For instance, the two complementary strands of DNA, each being a string of four kinds of chemical units, or bases – adenine (A), cytosine (C), guanine (G) and thymidine (T) – make some want to say that a C in one strand of the double helix will always line up with a G in the other, and an A is always paired with a T. Reductionism or scientific mechanism would like to find in this arrangement a template for the way that the information preserved by the base pairs will always be replicated. Usually it is true that there is a chemical affinity, which explains why C usually pairs with G, and A with T. "Sadly, real life is more complicated... heredity, the property that like begets like, depends not only on complementary base pairing, but on a complex dynamical system, involving both DNA and proteins."[191]

Within the Catholic theology of the Eucharist, certain traditions suggest that the bread and wine transformed into the Eucharist need to be accounted for in this conversion in a properly nuanced way. This alternative to Aristotelian scientific philosophy, and as adopted by St. Thomas Aquinas and his successors, would find its grounding in the form of the analogical imagination that informed the theology of the Greek Fathers of the Church in the fourth century CE. Theirs is a dynamic, even cosmic, theory of salvation history that sees in the Eucharist the bread and wine elevated to their highest meaning or potential.[192]

Western theology has its own version of this approach to the Eucharist. In Mark's Gospel, we find two descriptions of a storm at sea which Jesus calmed. After the first time that Jesus' word of command calmed the storm (Mark 4:35-41), his disciples asked themselves, "Who then is this, that even the wind and the sea obey him?" This scriptural account gave rise in the West to the theology of "obediential potency."[193] The core idea is that, within created nature, there is a radical openness to God and God's transformation of it through the Word. With respect to the bread and wine in the Eucharistic conversion, this tradition would find something within them that could obey the Eucharistic "Word of Command" so that they would no longer be bread and wine but would become the Body and Blood of Christ.

Like Berengar in the eleventh century, eucharistic theologians today try to answer the following questions for their own culture, which is essentially scientific in its mindset: (1) What are bread and wine? (2) What change takes place with the bread and wine? (3) How is the change to be conceived? (4) When does the change take place? (5) By what instrumental cause does the change take place?[194] Today, thanks to the science of genomics, we know a great deal more about actual bread and wine. As to what change actually takes place in the Eucharist, and when it happens, we would have to plead ignorance at the present time. However, the complexity of evolution and the replication of DNA do suggest that there might emerge answers in the future. As an analogy, could heredity be applied to the Divine Trinity or to the Second Person's relationship to the Eucharist? If it were so applied, would it not just be a matter of copying genes and stabilizing a dynamic system of interrelated replications? It is neither for three reasons: first, because of DNA; second, because of DNA's relationship to the enzymes; and third, because of the protein synthesis process.[195]

Perhaps the answer to the second question might emerge when we understand, in some future moment, what the so-called junk DNA or the DNA desert is. The February 2001 publication in *Nature* and *Science* of the first drafts of the human genetic code contained some surprises, one of the biggest being that we, as yet, do not understand the genetic instructions contained in 97 per cent of our DNA. The slogan "one gene, one enzyme" is a consoling one for those who are uncomfortable with complexity. We understand quite well the mechanism whereby "different triplets of bases in the DNA specify different amino acids. The DNA that carries the information also has sequences meaning 'start translating here' and 'end of protein'."[196] Between what we might want to call the proper genes, there are long stretches of this mysterious DNA. The mystery lies in the fact that they are not translated into proteins. Some small amount of this DNA has a regulatory function that we now know about, but most of it is puzzling. We simply do not know about its many functions. Could it be possible that in this large amount of DNA, the simplifiers would label as "junk" something within it, or the DNA itself, that would serve as an instrument of the "obediential potency" that the theologians, who maintain a strict separation between nature and grace, have hypothesized must exist in created nature?

The use of the word "mystery" in connection with junk DNA was done advisedly. Within the transubstantiation paradigm of the Eucharist, there is much mystery. One criticism of it would be that there is too much mystery. The subjectivity of the bread and wine is lost in the simple declaration that, in the Eucharist, the substance is now the Body and Blood of Christ and what we see of the bread and wine are mere "accidents." These "accidents" have their own molecular biological complexity. In other areas of Thomistic theology, we find the idea that divine grace elevates nature, but also respects the internal laws and being of both grace and nature. Would it make the act of faith in the power of God to convert bread and wine into the Body and Blood of Christ any less significant if we were to find an answer to the third question regarding our concept of the eucharistic change? Could we provide an image or icon in terms of the science of molecular biology? Does (or can) the consecration of the bread and wine change the genome of the bread and wine? If this were the case could, or should, this difference be studied in scientific research as the blood on the Shroud of Turin has been analyzed?

The theology of the decoding of the DNA molecule and the realization of its part in all of life creates other areas of speculation. It would return to the study of Christ, or Christology, and his genetic inheritance. Inasmuch as the Second Person of the Trinity took up the DNA and genetic information of Jesus as the primary and unique expression of the Word, one can then ask how the glorified and risen Christ converts and transforms the DNA and genetic information of bread and wine in the Sacrament of the Eucharist so that it too can share in this divine-human reality.

Further questions might be asked about the risen body of Jesus Christ. Can the divine become experiential at all? If the answer is yes, we must ask how. What form would the risen body of Jesus and the assumed body of Mary take? Would DNA be involved?

According to scholastic theology, the human soul is the real form of the body. Is there any room left for the soul in the new science of DNA and genetics?

Our faith and our scientific rationality are collaborators, not opponents. Since Fr. Gregor Mendel's time, and especially in the last fifty years, we have learned a great deal about how the nature of bread and wine can be transformed through additional genetic information. What we do not know or understand is how to translate this new knowledge into the Catholic teaching on the Eucharist, understood as a divine transformation of bread and wine through the Word of God within the Church.

With regard to the overall discussion of Roman Catholic perspectives on transgenic modification of food, the following hypothesis is offered. Because of their theological reflection on the change of substance in the Eucharist from bread and wine into the Body and Blood of Jesus Christ, many Catholics might not find the possibilities offered today by biotechnology to be against nature or intrinsically unethical. The Catholic analogical imagination, shaped as it has been by the dogma of the real presence of Christ in the Eucharist, as articulated by the Council of Trent, has been accustomed to think about and accept into one's body as "true food...and true drink" (John 6:55) the Body and Blood of Christ, something far more radical than anything currently offered by biotechnology today.

Beginning with a distinction between two conceptual languages, analogical and dialectical, this part of the book has discussed theological discourse as used in the substantial equivalence debate over the regulatory methods to be employed for GMO technology and the discourse of molecular biology to see whether they might offer some potential for Catholic theological discourse around the "doing" of the Eucharist. It has found the first wanting, but argues that the second, genetic modification, might offer some potential for the contemporary articulation of the analogical imagination, possibly replacing the ancient Aristotelian and Thomistic categories.

Conclusion

Catholic perspectives on transgenically modified food depend upon which Catholic you ask. The farmer growing genetically modified herbicide-resistant canola will have a different point of view from anti-globalization protesters. The fact that they are both Catholics will not matter in their argument. Each will have a valid perspective; each will marshal arguments to persuade a third party that their point of view is the correct one.

There are few specific moral issues on which official Church teaching has taken a clear position that will not find some Catholics in dissent. On the question of GMOs, no clear position has been taken by the pope and the other bishops of the Church beyond cautious support for the new biotechnology buttressed by various caveats. Therefore, Catholics can take any point of view they want on this matter. If we move from the perspectives of individual Catholics to the way the Catholic community as a whole through history has dealt with new technologies, we find the same diverse approach.

The present book has attempted to explore Catholic tradition in ways that are intended to help resolve the present argument by going beyond (or beneath) vehement rejection of GMOs or their enthusiastic support. This has been done because a certain polarization has occurred, which means that further dialogue on the question has apparently come to a temporary halt.

In the first chapter on technology, the central point was that all technology is subject to sinful, anti-life applications. These dark uses of any breakthrough discovery are often inherent in the very technique, but were not averted to by the inventors or early proponents. Therefore it is suggested to those who enthusiastically support transgenic modification of food that there are problems with what we will choose to do with it, and these will become clear to us in time. To those who are opposed to biotechnology, the point to stress is that much depends on the people who will profitably use it and the way that they will choose life, or not.

The same "yes, but" approach appears in the second chapter on intellectual property rights. There are many problems with patents, and these can harm the best interests of consumers, especially in the developing world. However, there are growing indications that socialization is taking place so that we will be able to protect the best interests of all the stakeholders involved. Who would have imagined a few years ago that 39 of the world's largest pharmaceutical companies would choose, for the best interests of their stockholders, to withdraw their lawsuit against the apparent infringement of their patent rights by a new South African law aimed at providing affordable generic drugs in Africa?

Perhaps the hardest issue to face squarely is that of risks. In this question, how one chooses to interpret the available evidence depends upon one's perspective and point of view. Nevertheless, Catholics as a group tend to follow a middle course on this matter, one that is neither lax nor rigorous.

Finally, it is the basic thesis of the book that, thanks to the doctrine of the conversion or transformation of bread and wine into the Body and Blood of Christ in the Eucharist, Catholics can accept the process of transgenic modification, and even use some of the technical scientific discourse underlying it to discuss the deep meaning of their own beliefs.

For many Catholics, therefore, the critical question is one not of process but outcome. Who benefits from these discoveries, and are the benefits justly distributed? Who stands to lose? Who takes on the risks, and have they been given enough information about them and a real choice in this matter?

In medical ethics the individual's right to informed consent in treatment and research on human subjects, based on the paramount dignity of the human person, is a normative principle that most Catholics can endorse. Similarly, the obligation of producers to label their GMO products clearly for consumers would be a second norm that most would support.

Labelling of GMOs is an ethical necessity. The reader might then ask whether the author of this book would consume such products. The answer is affirmative. Furthermore, I hope that research and development of this new form of molecular biology will be focused on the needs of the poor for whom the struggle to find sufficient food and vitamins, as well as vaccines, is a daily and often dramatic one.

NOTES

[1]   Kevin Cox, Activist guilty in whipped-cream assault on PM, *Globe and Mail*, April 4, 2001, A3.

[2]   *Globe and Mail,* August 18, 2000, A6.

[3]   Anne McIlroy, Canadians wary of genetically altered foods, *Globe and Mail*, January 15, 2000, A2. This same survey also indicates a growing problem with consumer acceptance of gene-spliced food in the USA, whose population had been basically supportive in surveys done in the early 1990s; see Thomas J. Hoban, Trends in consumer attitudes about biotechnology, *Journal of Food Distribution Research* 27 (1996):1–10.

[4]   The Politics of Genes. America's next ethical war. *The Economist*, April 14, 2001, 21.

[5]   This is due to the process of "co-mingling," the mixing of genetically enhanced crops with traditional, non-transgetic material at processing plants where they are pressed into oils or ground into flour or meal. This practice, encouraged by the biotechnology industry and government regulators in Canada, now has deleterious effects on international trade; see, for example, The future of food is here. *National Post*, May 15, 1999, B5. The reason why GM food is so prevalent in Canada and the US has to do with the widespread use of Bt (Bacillus thuringiensis), a natural soil bacterium, in the bioengineering of corn and other crops. Corn is widely used in cooking oil, and its derivatives appear in most processed foods in the form of sweeteners and thickeners.

6   Discorso di Giovanni Pauolo II ai partegipanti ad in Convegno su Ambiente e Salute, *L'Osservatore Romano,* 24 marzo 1997; Jubilee of the Agricultural World, Address by John Paul II, Saturday, 11 November 2000.

7   G. Ancora et alii, *Biotecnologie Animali et Vegetali. Nuove frontiere e nuove responsabilità.* (Città del Vaticano: Libreria Editrice Vaticana, 1999).

8   GMO Roundup, *Nature Biotechnology* Vol. 18 (January 2000): 7.

9   Robin Marantz Henig, *The Monk in the Garden* (New York: Houghton Mifflin, 2000), 21–25.

10   *Ibid.,* 15.

11   *Ibid.,* 89–91.

12   For Canadians, two exceptions would be fiddleheads and wild blueberries, foods we commonly eat but which have not been genetically modified.

13   David T. Dennis, Why GM foods aren't so scary, *Financial Post,* October 14, 1999, C7.

14   David Tracy points out that in the Western tradition of moral philosophy one finds a variety of public ways to discuss policy issues. "[T]here seems little doubt that all ethical arguments are in principle open to all intelligent, reasonable and responsible persons," *The Analogical Imagination: Christian Theology and the Culture of Pluralism.* (New York: Crossroads, 1981), 9.

15   The word "subliminal" is used advisedly because research shows that just over 50 per cent of Catholics in the USA oppose biotechnology's involvement with the food industry. At the same time, this same survey showed that a majority believe that humans should use their knowledge to improve the world. These results are similar to the views of Protestants, Jews and Muslims studied at the same time; cf. The Pew Initiative on Food and Biotechnology and Genetically Modifying Food: Playing God or Doing God's Work? Survey conducted between July 16 to July 20, 2001, of 301 Catholic adults aged 18 and older by Zogby International in Utica, NY, on behalf of the *The Pew Charitable Trusts.*

16   Homicide is condemned in many places in the "covenant code," including Exodus 21:12-14. Clear condemnation of technology is harder to find. There are those who would cite the judgment of God against the people of Babel in Genesis 11 as an argument against technology

as such, but this interpretation is far from our understanding of the problem under discussion, and probably has nothing to do with the intention of the sacred authors.

[17] In recent years, 120 governments have negotiated a treaty known as the "Stockholm Convention on Persistent Organic Pollutants." This treaty attempts to curtail the use of twelve chemicals, known as the "dirty dozen," which have been linked to cancer and to birth defects. DDT is one of these, but see Amir Attaran, DDT saves lives, *Globe and Mail*, December 5, 2000; Sarah Boseley, Millions more malaria deaths feared if DDT is banned, *Guardian Weekly,* September 2–8, 1999, 3; DDT: A useful poison, *The Economist,* December 16, 2000, 90; Let's use DDT, *National Post*, August 19, 2000, A15.

[18] Witold Rybczynski, *One Good Turn: A Natural History of the Screwdriver and Screw.* (New York: Scribner, 1986); Ken Lamb, *P.L.: Inventor of the Robertson Screw.* (Milton, ON: Milton Historical Society, 1998).

[19] Lamb, *P.L., Inventor of the Robertson Screw,* 20–30.

[20] I am indebted to "Barbed Wire," an article by Reviel Netz (*London Review of Books*, July 20, 2000, 30ff). Although I will follow the basic history that Netz has presented, I will find in it a different significance.

[21] Pope John Paul II, *On Social Concerns,* #42 [emphasis his].

[22] Ursula Franklin, *The Real World of Technology.* CBC Massey Lecture Series. (Concord, ON: Anansi, 1990), 18–20.

[23] Henry D. and Frances T. McCallum, *The Wire that Fenced the West* (Norman, OK: University of Oklahoma Press, 1965), 37–39, 45–46; Jack Glover, *The Barbed Wire Bible IX* (Sunset, TX: Cow Puddle Press, 1996), www.rushcounty.org/BarbedWireMuseum/Bwhistory.htm

[24] McCallum, *The Wire that Fenced the West.*

[25] People and Discoveries. Watson and Crick describe structure of DNA 1953. http://www.pbs.org/wgbh/aso/databank/entries/do53dn.htm. James Watson had worse things to say about "Rosy," which he concluded by describing her future at King's College and in Wilkins' laboratory as follows: "Clearly Rosy had to go or be put in her place..."; James D. Watson, *The Double Helix. A Personal Account of the Discovery of DNA.* (New York: New American Library, 1969), 20. Anne Sayre puts the situation this way: "...although Rosalind came very close between January 1951 and March 1953 to solving the DNA problem,

she was beaten in the end by Crick and Watson, who had in their success more help from her work than she ever knew they had received," *Rosalind Franklin and DNA* (New York: W.W. Norton, 1975), 116.

[26] People and Discoveries. Rosalind Franklin 1920–58, *op. cit.*

[27] If the research into novel foods and genetic modification were done mainly in publicly funded universities, such as Cambridge University where Watson and Crick worked, or at non-profit organizations like the Rockefeller Foundation, it would not be necessary to patent the results.

[28] Natalie M. Derzko, Plant Breeders' Rights in Canada and Abroad: What Are These Rights and How Much Must Society Pay for Them? *McGill Law Journal* 39 (1994):153–4.

[29] Ted Schrecker, Harvard Mouse: Sound law, thoughtless policy, *Globe and Mail*, August 15, 2000, A13.

[30] James Podgers, Patent decision fuels genetic research debate, *American Bar Association Journal* (1980); cited in Winifride Prestwich, Gene patents' key step toward the fabrication of man, *The Interim*. April, 2000, 14.

[31] Anthony Ramirez, *New York Times*, May 14, 2000, 5; EPO admits patent mistake, *Nature Biotechnology* Vol. 18 (April 2000): 366.

[32] Alan McHughen, *Pandora's Picnic Basket. The Potential and Hazards of Genetically Modified Food*. (New York: Oxford, 2000), 246.

[33] U.S. Patent and Trademark Office (Washington); cited in *Harper's Index*, October 1999, 13, 95.

[34] James Meek, Attempt to "patent life" may double cost of breast cancer checks, *Guardian Weekly*, January 20–26, 2000, 9.

[35] *Ibid.*

[36] Derzko, Plant Breeders' Rights in Canada, 176.

[37] Justin Gillis, Monsanto offers patent waiver on "golden rice" to aid poor, *Guardian Weekly*, August 10–16, 2000, 28.

[38] McHughen, *Pandora's Picnic Basket*, 37–40.

[39] Jennifer Kahn, The Green Machine, *Harper's*, April 1999, 70.

[40] Canola farmer loses in Monsanto lawsuit, *Winnipeg Free Press*, March 30, 2001, B1.

[41] Mary Ambrose, "Percy Schmeiser: Seeds of Doubt," *Business FT*, Weekend Magazine, April 14, 1999.

[42] Cathryn Atkinson, Battlefields in a war against the giant, *Guardian Weekly,* February 10–16, 2000, 24.

[43] Adam Killick, Farmer's Battle with biotech giant Monsanto begins in Federal Court, *National Post*, June 6, 2000, A19.

[44] Ed White, "Farmer takes stand in patent-breaking lawsuit," *Western Producer,* June 22, 2000.

[45] Atkinson, "Battlefields," *Ibid.*

[46] Estimates of the ratio of farmers in India who replant saved seed each year vary. In 1990 only 10 per cent of farmers bought new seed each year; by 1997 this may have risen to 25 per cent; Vadana Shiva and Tom Crompton, Monopoly and Monoculture, *Economic and Political Weekly* XXXIII, No. 39, September 26–October 3, 1998, A138.

[47] John Vidal, The Seeds of Wrath, *Guardian Weekly,* July 8–14, 1999.

[48] In September 2001, RAFI changed its name to ETC, pronounced "et cetera," and functioning as an acronym for the Action Group on Erosion, Technology and Concentration.

[49] Between 1986 and 1994, during the Uruguay Round of trade talks, the member countries of the World Trade Organization, which includes India, worked out an agreement on Trade-Related Aspects of Intellectual Property Rights, also known as TRIPS. It was a framework that left to signatories the obligation to develop their own laws. In a parallel development, the International Union for the Protection of New Varieties of Plants (UPOV) reached its latest agreement in 1991. One battle within India was whether to draft legislation within the framework of UPOV or not. The UPOV system gives many more rights to a plant breeder like Monsanto than to farmers such as those in Warangal.

[50] "Terminator" fell between two pieces of legislation, one updating a 1970 patent law and the other on biodiversity. Both died when the coalition government fell in 1999 and a new election was conducted in India; cf. S. Gopikrishna Warrier, *Business Line,* May 20, 1999.

[51] A.A. Berle, Jr., *Power Without Property* (New York: Harcourt, Brace, 1959), 60; as cited in E. Duff, Property, Private, *New Catholic Encyclopedia,* Vol. XI (New York: McGraw-Hill, 1967), 849.

[52] John Paul II, *Sollicitudo Rei Socialis*, no. 9, Washington, DC: United States Catholic Conference, 1988, 14–15; cf. also address of John Paul II, Jubilee of the Agricultural World, 11 November 2000.

[53] *Ibid., no.* 42.

[54] John XXIII, *Mater et Magistra*, no. 19; cited in Duff, Property, Private, *op. cit.*

[55] John XXIII, *Pacem in Terris*, April 11, 1963, no. 11; cited in Duff, Property, Private, 850.

[56] Thomas Aquinas, *De reg. Princ.*, 1.15; Duff, Property, Private, *op. cit.*

[57] Thomas Aquinas, *Summa Theologia* 2a2ae, 66.2; Duff, Property, Private, *op. cit.*

[58] Daniel Rush Finn, The economic personalism of John Paul II: Neither right nor left, *Journal of Scholarship for a Humane Economy* Volume 2, Number 1 (Spring 1999).

[59] Duff, Property, Private, 851.

[60] Much discussion of these scriptural texts took place in the year 2000 discussions on debt forgiveness, held by the G-7 industrialized nations and other similar organizations.

[61] It would be "rival," rather than "non-rival" when a country, competitor or individual who subsequently made use of your discovery had not participated in its original cost of creation, notwithstanding the fact that their enjoyment of it would not diminish its value to them or others. In addition to patents, an alternative to avoid the "rival" problem is secrecy, such as shrouded the recipe for Colonel Saunders' Kentucky Fried Chicken or the proprietary codes for Microsoft software. The possible theft in late 2000 of these codes indicates that every method of protecting intellectual property has its problems; cf. Worming out the truth, *The Economist,* November 4, 2000, 89.

[62] The original meaning of Matthew's parable may well be spiritual and have to do with sharing one's faith with others. In other words, the homilists use of this passage to recommend that we share our human gifts or ideas with others may not be based on sound exegesis of Matthew's text; cf. Leslie Brisman, A Parable of Talent, *Religion and the Arts* 1:1 (Fall 1995): 74–99.

[63] Denis Edwards, The ecological significance of God-language. *Theological Studies* 60:716.

[64] Bonaventure, *Hexaemeron* 12; cited in Edwards, The ecological significance, 717.

[65] *Ibid.*, 13.14.

[66] John Locke, *Concerning the True, Original Extent, and End of Civil Government*; cited in Robert W. Herdt, *Enclosing the Global Plant Genetic Commons*. The Rockefeller Foundation, 1999.

[67] Until recently, the same could have been said about the wind as about sunlight. The wind has become the fastest growing source of electricity in the USA; cf. Douglas Jehl, Curse of the Wind Turns to Farmers' Blessing, *New York Times*, November 26, 2000, 1, 32.

[68] Herdt, *Enclosing the Global Plant*.

[69] David Malakoff, Iridium accelerates squeeze on the spectrum. *Science* 282, Oct. 2, 1998, 34–5; cf. Herdt, *Enclosing the Global Plant*.

[70] *Ibid.*

[71] An important body dedicated to this task is the World Intellectual Property Organization (WIPO) of the United Nations. WIPO has its headquarters in Geneva, Switzerland, and has defined a patent as follows: "a patent is an exclusive right granted for an invention, which is a product or a process that provides a new way of doing something, or offers a new technical solution to a problem." It points out that this exclusive right is time bound, often for twenty years.

[72] G.J.M. Pearce, Augustine's Theory of Property, *Studia Patristica* 6 (1962): 498.

[73] Augustine, *Sermo,* 50, 2, 4. As cited in Barry Gordon, *The Economic Problem in Biblical and Patristic Thought* (Leiden: E.J. Brill, 1989), 124.

[74] D.J. MacQueen, St. Augustine's Concept of Property Ownership, *Recherches Augustiniennes* 8 (1972): 218, cited in Gordon, *The Economic Problem,* 124.

[75] Gordon, *The Economic Problem*, 125.

[76] T.F. Divine, Economic Value. *New Catholic Encyclopedia* Vol. 5 (New York: McGraw-Hill, 1967), 61.

[77] L. Geraci, Appunti sul tema del diritto morale di invenetore nel procedimento di brevettazioni. *Revista di Diritto Industriale* XXXII (1983), Parte Prima: 22–34; V. Buonomo, Brevetti e brevettabilità della biotechnologie: alcune consideratzioni sugli aspetti etici e giuridici, In: G. Ancora et alii, *Biotecnologie Animali e Vegetali. Nuove frontiere e*

nuove responsabilità. (Città de Vaticano: Libreria Editrice Vaticana, 1999); H.C. Hanson *International intellectual property law & policy.* 2 Vols. (London: Sweet & Maxwell, 1996). For instance, under the Trade Related Intellectual Property Rights (TRIPS) arrangements of the General Agreement on Trade and Tariffs (GATT), signatories are required to confer legally exclusive property rights to seed producers. If these are conferred through legal systems consistent with the International Union for the Protection of New Varieties of Plants 1991, farmers will be divested of their rights to exchange seed, and even to save seed for their own use; cf. Shiva and Crompton, Monopoly and Monoculture, *op. cit.* A141.

78 Ian MacLeod, Biotech companies seek to patent gene data, *National Post*, April 5, 2001, A10.

79 http:www.life.ca/nl/70/terminator.html

80 John Paul II, *Sollicitudo Rei Socialis*, no. 38.

81 *Ibid.*, Address of November 11, 2000, Jubilee of the Agricultural World, emphasis in the original text – http://www.vatican.va/holy_father/ john_paul_ii/speeches/2000/oct-dec/documents/hf_jp-ii_spe_20001111_jubilagric_en.html

82 David M. Gould, William C. Gruben, The role of intellectual property rights in economic growth, *Journal of Development Economics* 48 (1996): 323.

83 Gene M. Grossman and Elhanan Helpman, *Innovation and Growth in the Global Economy* (Cambridge, MA: MIT Press, 1991); Paul M. Romer, Endogenous growth and technical change, *Journal of Political Economy* 98:71–102; cited in Gould and Gruben, The role of intellectual property. *op. cit.,* 323.

84 This was the April 10, 2000, decision taken by the Human Genome Organization, which represents members in 57 countries; Carolyn Abraham, *Globe and Mail,* April 11, 2000, A6.

85 For instance, patents are applied and granted for "the smallest fragments of genetic material...[such as] short fragments of DNA used as genome markers like expressed sequence tags (ESTs) and single nucleotide polymorphisms (SNPs) [which], although they may not directly identify genes, may still be extremely useful and thus satisfy the utility requirement and hence be patentable." Robert W. Herdt, *Enclosing the Global Plant Genetic Commons, op. cit.*

[86] Rainer Moufang, Methods of Medical Treatment Under Patent Law, *International Review of Industrial Property and Copyright Law* 24 (1993):18.

[87] http://www.merck.com/overview/philanthropy/mectizan/p12.htm

[88] In a separate development possibly related to the Merck mission statement, in March 2001 the company announced that it had negotiated a discount price for Brazil of its anti-AIDS drug Stocrin (efavirenz) in the face of the Brazilian government's threat to start manufacturing the drug itself. "We want to be able to provide Brazil with medicines at an affordable price," said Gregory Reaves, a Merck spokesman. "We want to resolve this matter in discussions"; Patents, profits, and AIDS care. *Monitor World*, March 12–18, 2001, 4. In a similar manner and in the same week, Merck announced that it was cutting the price in Africa of two of its AIDS-fighting drugs to a tenth of what the drug costs in America. Crixivan (indinavir suphate), which costs $6,000 a year in the United States, is on offer to sub-Saharan Africa at $600 a year, and Stocrin at $500. Merck says it will make no profit on the sale; cf. A war over drugs and patents, *The Economist*, March 10, 2001, 43; Sarah Bosely, Embarrassed firms slash prices for AIDS drugs. *Guardian Weekly,* March 15–21, 2001, 4.

[89] Donald G. McNeil Jr., New York Times Service, *International Herald Tribune,* May 23, 2000.

[90] Naomi Koppel, Associated Press Online, Feb. 9, 2001.

[91] In Brief, *Guardian Weekly*, March 8–14, 2001, 12.

[92] J. Madeleine Nash, Grains of Hope, *Time*, July 31, 2000, 17; Aaron J. Bouchie, Golden handouts on the way, *Nature Biotechnology* Vol. 18, (September 2000): 911.

[93] *Ibid.*

[94] Justin Gillis, Monsanto offers patent waiver on "golden rice" to aid poor, *Guardian Weekly*, August 10–16, 2000.

[95] Naomi Klein, There's nothing like a feel-good bowl of golden rice. Or not. *Globe and Mail*, August 2, 2000, A13.

[96] J.B. Walker, Antoninus, St., *New Catholic Encyclopdedia,* Vol. 1 (New York: McGraw-Hill, 1967), 647.

[97] John T. Noonan, Jr., *The Scholastic Analysis of Usury* (Cambridge, MA, 1957), 78. The enormity and insidiousness of usury was impressed upon St. Antoninus by his daily contact with it in the highly commer-

cialized society in Florence. Usury, according to him, is the great harlot of Apocalypse 17, "Who sitteth upon many waters, With whom the kings of the earth have committed fornication" (*Summa moralis* 2.1.6).

[98]   Antoninus of Florence, *Summa Moralis* I.1,3,ii.

[99]   David Hollenbach, *Claims in Conflict: Retrieving and Renewing the Catholic Human Rights Tradition* (New York: Paulist, 1979), 55.

[100]   At the 50th Anniversary Assembly of the World Council of Churches, its eighth, held in Harare, Zimbabwe, in December 1998, the delegates approved the following declaration: "It is our deep conviction that the challenge of globalization should become a central emphasis of the work of the WCC, building upon many significant efforts.... The vision behind globalization includes a competing vision to the Christian commitment to the oikoumene, the unity of human kind and the whole inhabited earth.... We should not subject ourselves to the vision behind [globalization], but strengthen our alternative ways towards visible unity in diversity, towards an oikoumene of faith and solidarity." The Eighth Assembly left the details of how this opposition to globalization should relate to GMOs to the permanent "justice, peace and creation committee" of the WCC.

[101]   The Royal Society of Canada, *Elements of Precaution: Recommendations for the Regulation of Food Biotechnology in Canada*. An Expert Panel Report on the Future of Food Biotechnology prepared by the Royal Society of Canada at the request of Health Canada, Canadian Food Inspection Agency and Environment Canada, January 2001, 20. The Royal Society report cited a 1997 study by Industry Canada to the effect that the global market for biotechnology products will reach $50 billion annually by 2005 (Sector Competitiveness Frameworks. Bio-Industries: Part I Overview and Prospects, Bio-Industries Branch, Industry Sector, Industry Canada, March 1997).

[102]   Chris Cobb, Canadians wary of genetically modified foods, *National Post*, January 2, 2001, A6.

[103]   *Ibid.*

[104]   Anne McIlroy, Canadians wary of genetically altered foods, *Globe and Mail*, January 15, 2000, A2.

[105]   Kevin Rollason, Winnipeggers suspicious of modified food. Survey finds almost half of city harbours doubts about GM food, *Winnipeg Free Press*, February 19, 2001, A9.

[106] The biotech industry and some American politicians claim that genetically engineered "golden rice" would save the sight of 500,000 children a year. This is open to dispute. The Rockefeller Foundation is funding the rice's development. Gordon Conway, President of the Rockefeller Foundation, said: "We do not consider golden rice to be the solution to the Vitamin A deficiency problem. Rather it provides an excellent complement to fruits, vegetables and animal products in diets, and to various fortified foods and vitamin supplements." The environmental group, Greenpeace, is less polite. Its spokesperson said that a child in the Third World would have to eat about 16 pounds of cooked rice daily to obtain the benefits advertised. Greenpeace has complained to the Advertising Standards Council about this; cf. Paul Brown, GM rice promoters' claims "have gone too far." Adding vitamin A offers no quick fix to cure blindness, *Guardian Weekly*, February 15–21, 2001, 5; Clayton Ruby, Forget about labels, just eat what Ottawa puts in front of you, *Globe and Mail*, February 15, 2001, A11.

[107] Franken sense, *National Post*, January 24, 2000, A15.

[108] Judy Jamison, An edible hepatitis vaccine, *Nature Biotechnology*, Vol. 18 (November 2000): 1130.

[109] Clayton L. Rugh, Julie F. Senecoff, Richard B. Meagher, and Scott A. Merkle, Development of transgenic yellow poplar for mercury phytomediation, *Nature Biotechnology*, Vol. 16 (October 1998): 925–8.

[110] As will be explained further, "probabilism" in Catholic theology does not relate to situations like the fact that a US study in April 2001 found that there was only a 20 per cent chance that a product labelled "GMO-free" contained no genetically modified ingredients; cf. *Wall Street Journal* as cited in *Harper's Index,* October, 2001.

[111] Royal Society of Canada, *Elements of Precaution*, 225; Peter Calamai, Ottawa rapped in GM food report, *Toronto Star*, Feb. 5, 2001, 16.

[112] John Burgess, US backs down on GM food labelling, *Guardian Weekly*, February 3–9, 2000, 35.

[113] Gord Surgeoner, Genetically modified fries with that? *Globe and Mail,* January 24, 2000.

[114] Michael Wilson, Genetically Modified Tomatoes: Seeking Firmer Tomatoes with Better Flavour. In *Engineering Genesis. The Ethics of Genetic Engineering in Non-Human Species*. Ed. Donald Bruce and Ann Bruce. Society, Religion & Technology Project, Church of Scotland. (London: Earthscan Publications Ltd, 1999), 51.

[115]   J.L. Fox (1994). FDA Nears Approval of Calgene's Flavr Savr™, *Bio/Technology*, vol. 12, 439; cited in Wilson, Genetically Modified Tomatoes, 51–2.

[116]   Seeds of Discontent, *The Economist*, February 20, 1999, 75–7.

[117]   Masayoshi Kanabayashi, Japan's Ajinomoto stems damage in Indonesia over pig enzymes, *The Wall Street Journal*, February 1, 2001.

[118]   Malaysia seizes food seasoning from Indonesia, Associated Press, January 29, 2001.

[119]   Hugh Pennington, The English disease, *London Review of Books*, 14 December 2000, 3

[120]   Office International des Epizooties (Paris), cited in *Harper's Index*, March 2001, 15, 94.

[121]   Suzanne Daley, Rising rate of mad cow disease alarms Europe, *New York Times International*, May 7, 2000.

[122]   Hugh Pennington, The English disease, 3.

[123]   Hugh Pennington, The English disease, 5.

[124]   *The Inquiry into BSE and Variant CJD in the United Kingdom*, Nov. 20, 2000, Chapter 4, http://www.bseinquiry.gov.uk

[125]   A further experiment that issued the preliminary result that traces of BSE had been found in scrapie-infected sheep's brains has been dismissed. On November 30, 2001, it was announced that researchers in the Institute for Animal Health in Edinburgh had mixed up cow brains with sheep brains. DNA tests show that these scientists have spent the past three years analyzing cow brains when they thought they were dealing with sheep brains. This means that the scientific work on the vectors of BSE and its relationship to scrapie must begin again; Pallab Ghosh, Verdict awaited in BSE brains mix-up, BBC News, Friday, November 30, 2001, 04:18 GMT.

[126]   Hugh Pennington, The English disease, 6.

[127]   *Ibid.*

[128]   *Ibid.*

[129]   Government health officials in Belgium banned Coca-Cola in the spring of 1999 after a number of people who had drunk it complained of suffering nausea and vomiting. Coca-Cola attributed the problem to substandard carbon dioxide used to put bubbles in the drinks in one

plant, and a fungicide on wooden pallets used to transport cans at another plant. Earlier, the discovery that a factory in Belgium used fat laced with dioxins led the European Union to ban exports of Belgian products that might contain carcinogens; Food Scares Mount Up, Perplexing Europeans, *New York Times International,* Sunday, June 26, 1999.

[130] Royal Society of Canada, *Elements of Precaution,* 194.

[131] *Ibid.*

[132] Heather Scoffield, Canola operation could leave Canada over Europe's uproar, *Globe and Mail,* May 16, 2000, A10.

[133] David Cooper, Homeric Cheese v. Technophiliac Relish, *London Review of Books,* 18 May 2000, 28.

[134] This major worry has been alleviated by recent research. In a ten-year study published in *Nature,* Michael Crawley and his team at the Imperial College in London, England, found that, rather than being super plants, the GM crops in fact are unable to compete with natural ones without constant human support. Left to themselves in the wild, they are almost certain to die. The reason is that so-called natural crop plants are products of centuries of breeding by humans, but are also the result of natural selection. Nature seems reluctant to embrace genetic alterations and, untended by humans, the crops do not survive; Crawley, M.J., Brown, S.L., Hails, R.S., Kohn, D.D. & Rees, M. Transgenic crops in natural habitats. *Nature* 409 (2001): 682–3; Keep testing GMOs, *Winnipeg Free Press,* February 24, 2001, A18; Anne McIlroy, Engineered food crops don't spread, U.K. study finds, *Globe and Mail,* February 8, 2001; Genetically modified weaklings, *The Economist,* February 10, 2001, 79. However, the discovery that wayward genes of genetically modified corn have invaded up to 70 per cent of wild maize in remote mountainous regions of Mexico, as reported by Ignacio Chapello in *Nature,* November 2001, gives cause for real concern. "It's like an epidemic," says Chapello. Margaret Munro, Wild corn tainted with genes from engineered species, *National Post,* November 29, 2001, A2.

[135] *Ibid.*

[136] Royal Society of Canada, *Elements of Precaution,* 198.

[137] David Cooper, Homeric Cheese, 29; B. Honings, La persona umana al centro della natura e della scienza, *Biotecnologie animale e vegetali. Nuove frontiere e nuove responsabilità.* (Vatican City: Libreria Editrice, 1999),

60–72; A. Pessina, Note sul rapporto tra biotecnologie e antropologia filosofica, 73–82.

[138] D. Gayle Johnson, Facts, trends and issues of open markets and food security, International Agricultural Economics Association workshop, Berlin, August 12, 2000. With regard to the possibility that Western risk aversion to the further development of GM nutritional supplements in common food such as rice will result in discontinuing research on them, Johnson says: "But the important issue is…the lost opportunities for improving the nutritional status of poor people in developing countries. Is anyone who promulgates the anti-GMO hysteria thinking about this? Or do they care? Perhaps the problem is that they don't care about what the consequences would be for others less fortunate than they are." (14.)

[139] Albert R. Jonsen and Stephen Toulmin, *The Abuse of Casuistry. A History of Moral Reasoning.* (Berkeley: University of California Press, 1988), 164–5.

[140] Jonsen and Toulmin, *Abuse of Casuistry*, 166.

[141] *Ibid.,* 174.

[142] Germain Grisez, *The Way of the Lord Jesus. Christian Moral Principles* Volume 1 (Chicago: Franciscan Herald Press, 1983), 12.D, 292; DS 2101-67/1151–1216. A seventeenth-century example of Laxism is the case of a nobleman's servant who helped his master violate a virgin by holding the ladder while the nobleman climbed through her window. The Laxists would say that under certain circumstances, such co-operation was permissible.

[143] Grisez, *The Way of the Lord Jesus,* 292; DS 2302/1293.

[144] For example, with regard to the moral status of the early embryo, the official teaching of the Church's magisterium would seem to be, notwithstanding the 1690 decision, that of "tutiorism," inasmuch as full personhood is bestowed on the newly conceived zygote even though the philosophical question of ensoulment, especially before the fourteenth day after conception, remains undecided.

[145] http://www.wcc-coe.org/wcc/what/jpc/ecology.html

[146] David Tracy, *The Analogical Imagination* (New York: Crossroad, 1981), 408.

[147] Tracy, *The Analogical Imagination,* 413, as cited in Mark S. Massa, Anti-Catholicism and the Analogical Imagination, *Theological Studies* 62:3, 564.

[148] Tracy, Analogical Imagination, 415.

[149] Elizabeth Levine, Sites of DNA-Altered Wheat Secret: Ottawa, *National Post*, July 31, 2001, A4.

[150] Elizabeth Levine, Wheat coalition urges debate on gene modifying, *National Post*, August 1, 2001.

[151] National Conference of Catholic Bishops, "The Real Presence of Jesus Christ in the Sacrament of the Eucharist: Basic Questions and Answers," *Origins. CNS Documentary Service.* Volume 31, No. 7 (June 28, 2001), 124.

[152] For example, Tibor Horvath, *Eternity and Eternal Life* (Waterloo, ON: Wilfrid Laurier University Press, 1993).

[153] Claus Emmeche, Autopoietic Systems, Replicators, and the Search for a Meaningful Biologic Definition of Life. *Ultimate Reality and Meaning* 20: 244–64.

[154] Karl Rahner, *Foundations of Christian Faith.* Trans. William V. Dych, (New York: Seabury Press, 1978), pp. 65–6, 119.

[155] First Vatican Council, Faith and Reason, Chapter 4; in *Decrees of the Ecumenical Councils,* Vol. 2, Edited Norman P. Tanner (Washington, DC: Georgetown University Press, 1990), 808–9; *Faith and Reason. Encyclical Letter Fides et Ratio of the Supreme Pontiff John Paul II on the Relationship between Faith and Reason.* Città del Vaticano, 1998.

[156] In what follows, I am grateful to Professor Tibor Horvath SJ for permission to use some of his reflections on Aquinas and the Eucharist from his (as yet) unpublished manuscript entitled "Thinking about Our Faith."

[157] First Vatican Council, *Faith and Reason,* cf. Tanner, 1990, 808.

[158] Augustine's statement was: "*Accredit verbum ad elementum et fit sacramentum*; The word is added to the element and becomes a sacrament"; *Tractatus super Joannem,* 80.3; PL 35:1840; ST3.78.5; cf. Egil Grislis, The Eucharistic Presence of Christ: Losses and Gains of the Insights of St. Thomas Aquinas in the Age of Reformation. *Consensus. A Canadian Lutheran Journal of Theology* 18:15, 31.

[159] Evelyn Fox Keller, *The Century of the Gene.* (Cambridge, MA: Harvard University Press, 2000), 1–3; John Maynard Smith, "The Cheshire Cat's DNA," *The New York Review of Books*, Vol. XLVII, Number 20, December 21, 2000, 443.

160 Geoffrey Carr, "The Human Genome," *The Economist*, July 1, 2000, 4.

161 Alan McHughen, *Pandora's Picnic Basket,* 21–2.

162 There are a few cell types lacking DNA, most notably red blood cells; *Ibid.*, 22.

163 *Ibid., 23.*

164 Carr, "The Human Genome," 4.

165 *Ibid., 5.*

166 *Ibid.*, 5.

167 Royal Society of Canada, *Elements of Precaution: Recommendations for the Regulation of Food Biotechnology in Canada*, January 2001, 15.

168 The situation is somewhat different with evolutionary biology. In contrast to the methods of biotechnology, evolutionary biology does not make this change "suddenly," as in the substantial and radical change that happens to the bread and wine in the Eucharist, but gradually over time. For instance, we understand that the human species has built upon the evolutionary history of primate history. Because we share most of the primate DNA sequences, many scientists would be reluctant to describe the change from primate to human as "substantial," and would similarly argue that the change from maize to Bt Corn is not substantial. This is at the heart of an important debate.

169 Michael Specter, "The Pharmageddon Riddle," *The New Yorker*, April 10, 2000, 63.

170 Laurent Belsie, Genetically modified corn broadens public backlash, *The Christian Science Monitor/MONITORWORLD*, Nov. 20–26, 2000, 6.

171 *The New York Times National*, Sunday, October 22, 2000.

172 Stuart Laidlaw, Taco shell turmoil spills into Canada. Consumers here not being told about GM corn, *Toronto Star,* Oct. 26, 2000, B8; James Baxter, Bio-tech corn not in Canada, feds say, *Calgary Herald*, October 27, 2000, A8.

173 Statement by Stephen Johnson, EPA Assistant Administrator for Pesticides, Environmental Protection Agency, Washington DC, October 12, 2000.

174 Letter of Sally Van Wert, Regulatory Affairs – Biotechnology, Aventis CropScience USA, November 22, 2000, addressed to Mr. Paul Lewis, EPA, Washington, DC, USA.

[175] The concept of "substantial," as used in bioengineering, would not have been called this by scholastic philosophers because it has to do with "accidents" or "the accidental." This means that it has to do with the experiential,the measurable, and *not* with the substantial. The consecrated bread experienced scientifically is only bread. The notion of substance in science is not the same as the Thomistic sense of the word.

[176] Royal Society of Canada, *Elements of Precaution*, 177.

[177] Guidelines for the Safety Assessment of Novel Foods, Volume I, Food Directorate, Health Protection Branch, Health Canada, September 1994.

[178] Steve Taylor, Professor and Head, Food Science and Technology Department, University of Nebraska, Susan Hefle, Co-Director, Food Allergy Research and Resource Program, University of Nebraska, "Response to 'Beyond Substantial Equivalence'."

[179] Biotechnology Industry Organization and Grocery Manufacturers of America, October 9, 1999, http://www.bio.org

[180] E. Millstone, E. Brunner and S. Mayer, 1999. "Beyond substantial equivalence." Nature 401:525–6.

[181] Genes and pagadigm shiftiness. Prince Charles and Roy Romanow could use a translator, *Globe and Mail*, June 3, 2000, A14.

[182] The fact that we can purify the DNA of a bacterial toxin from the cells of genetically modified corn and remove the Bt modification, known to molecular biologists as a genetic revertant, thereby recovering the "original" plant variety, is a further argument against the development of a strict analogy between bioengineering and the Catholic understanding of what happens at the Eucharist because there is no way that the consecrated species can be returned to simple bread and wine.

[183] Cat. 22, Mystagogica 4, 1. 3–6. 9: PG 33, 1098–1106

[184] Edward J. Kilmartin, SJ, *The Eucharist in the West. History and Theology.* Edited Robert J. Daly, SJ (Collegeville, MN: Liturgical Press, 1998), 98.

[185] PL 150.66B; Kilmartin, *op. cit.*

[186] Hans Grass, *Die Abendmahlshere bei Luther und Calvin* (Gutersloh: C. Bertelsmann, 1954), 127; WA 26:447:14ff; LW 37:306.

[187] Council of Trent, Session 13, canons 2 and 3; cited in Edward J. Kilmartin, SJ, *The Eucharist in the West: History and Theology.* Edited Robert J. Daly, SJ (Collegeville, MN: The Liturgical Press, 1998), 180.

[188] John C. Haughey, SJ, "Connecting Vatican II's Call to Holiness with Public Life," *Catholic Theological Society of America Proceedings* 55/2000, 11.

[189] In his study of *"gratia operans,"* Bernard J. F. Lonergan in *Grace and Freedom. Operative Grace in the Thought of St. Thomas Aquinas*, Vol. 1, *Collected Works*. Ed. Frederick E. Crowe and Robert M. Doran. (Toronto: University of Toronto Press, 2000), 26–7 called the distinction between the purely natural and the supernatural a "theory" proposed in the thirteenth century by Philip the Chancellor, who "presented the theory of two orders, entitatively disproportionate…"

[190] Kilmartin, *The Eucharist in the West*, 204.

[191] Smith, The Cheshire Cat's DNA, 43; Keller, *Century,* 55–72.

[192] Kilmartin, *The Eucharist in the West*, 150, Gregory of Nyssa, *The Great Catechism*, Chapter XXXVII.

[193] The context of the development of the idea of "obediential potency" in the West was the problem that the miraculous posed for the theory of the radical separation between nature and grace. First appearing in the *Summa* attributed to Alexander of Hales, *potentia obedientialis* explained why the miraculous did not actually provide data proving that the separation between the two orders was only partial. These exceptions seemed to cast theological doubt upon the "theory of the intrinsically supernatural" but had to be explained within the terms of reference of the theorem itself.

[194] Kilmartin, *The Eucharist in the West*, 98.

[195] Keller, *Century,* 54–5; Smith, The Cheshire Cat's DNA, 43.

[196] Smith, The Cheshire Cat's DNA, 44.

AGMV Marquis

MEMBER OF SCABRINI MEDIA

Quebec, Canada
2002